Continuous Dislocation Dynamics Made Simple

Online at: https://doi.org/10.1088/978-0-7503-6330-3

Continuous Dislocation Dynamics Made Simple

Alfonso Ngan
Alireza Kalaei
University of Hong Kong, Hong Kong, People's Republic of China

IOP Publishing, Bristol, UK

ISBN 978-0-7503-6330-3 (ebook)
ISBN 978-0-7503-6328-0 (print)
ISBN 978-0-7503-6331-0 (myPrint)
ISBN 978-0-7503-6329-7 (mobi)

DOI 10.1088/978-0-7503-6330-3

Version: 20240701

IOP ebooks

British Library Cataloguing-in-Publication Data: A catalogue record for this book is available from the British Library.

Published by IOP Publishing, wholly owned by The Institute of Physics, London

IOP Publishing, No.2 The Distillery, Glassfields, Avon Street, Bristol, BS2 0GR, UK

US Office: IOP Publishing, Inc., 190 North Independence Mall West, Suite 601, Philadelphia, PA 19106, USA

Contents

Foreword

This book on Continuous Dislocation Dynamics (CDD) by Alfonso Ngan and his student Alireza Kalaei is a particularly timely contribution to an important, evolving field. While the field of dislocation theory began over a century ago, its impact on computational mechanics has largely been heuristic. This is, in large part, the result of the need to account for a very large number (length/volume) of dislocations to appropriately reproduce the constitutive response of a plastically deforming material. This challenge is further compounded by the complex kinematics of dislocation lines in 3D, the long-range nature of their interactions and the intrinsic non-linearities associated with dislocation cores (their discrete, atomic nature and sensitivity to the nature of atomic bonding). These issues demand approaches which are quantitative, multi-scale, and capable of accounting for the complexity at levels appropriate for the plasticity questions under consideration. CDD is remarkable in its ability to accomplish much of this at a sufficiently reasonable computation cost to describe plastic deformation on the dislocation microstructure to macro-specimen scales by representing dislocations as a continuum density instead of discrete line defects. This book strikes a balance between physical description, theoretical development, and computational application and does so in a manner accessible to those new to the field and appropriate for experienced specialists. This balance and the necessary clarity of presentation in this book are a testament to the authors' understanding of the marriage of dislocation dynamics and continuum methods and their perspicuity.

David J Srolovitz
Chair of Materials Theory
Dean of Engineering
The University of Hong Kong

Preface

This book presents an easy-to-read introduction to continuous dislocation dynamics (CDD). By treating dislocations, which are carriers of crystal deformation as continuously spread defects represented by a density, CDD aims at filling the mesoscopic gap in multi-scale materials modelling. CDD is more coarse-grained than discrete dislocation dynamics (DDD), which models dislocations as discrete lines represented by nodes, and is less coarse-grained than slip-system based FEM crystal plasticity (CPFEM), which models plastic deformation as resulting from slip of crystal planes without important dislocation physics incorporated. Therefore, compared with DDD, CDD can treat much bigger problems involving real quantities of dislocations in engineering applications, yet compared with CPFEM, key dislocation physics are preserved.

This book aims at introducing CDD as an exact, rather than approximate or ad hoc, approach to crystal plasticity. The starting point will be Maxwell's equation for moving lines, which will be generalized to densities of parallel dislocation quantities, and then to an 'all-dislocation' density (ADD) defined on both the real-space and dislocation-character domains. The ADD is a representation of dislocation density of a coarse-grained resolution suitable for handling large systems, and yet an exact kinematics law is available. This aspect is not yet well understood/appreciated by the research community and hence this book aims to make this a focus.

Dislocation dynamics, given its involved mechanics and mathematical foundation, is never a simple subject. While most existing models for dislocation kinematics are cast in rather complex equations, which at the same time may not be exact, central to this book is a recently derived exact kinematics law for the 'all-dislocation' density that is also minimalist in mathematical form. This book is written in a style aiming to facilitate understanding of the difficult mathematical concepts of dislocation dynamics, by providing simple illustrations and key computer codes. The aim is that a senior undergraduate or starting graduate student should find the book readable. The reader is assumed to have basic understanding of dislocations, as covered in most undergraduate textbooks.

Alfonso H W Ngan
Alireza Kalaei
May 2024, Hong Kong

Author biographies

Alfonso Ngan

Alfonso Ngan is currently Kingboard Professor in Materials Engineering and Chair in Materials Science and Engineering at the University of Hong Kong (HKU). He obtained his PhD from the University of Birmingham, UK in 1992, and carried out postdoctoral research work at the University of Oxford, before joining HKU in 1993. His research interests include nanomechanics, smart actuating materials, and microstructural basis of material properties. His research-related honours include Fellow of the Royal Academy of Engineering and the Hong Kong Academy of Engineering Sciences, Guanghua Engineering Science and Technology Prize by the Chinese Academy of Engineering, the Rosenhain Medal and Prize from the Institute of Materials, Minerals and Mining, higher doctorate (DSc) from the University of Birmingham, and the Croucher Senior Research Fellowship. He is currently an Associate Editor of the International Journal of Plasticity.

Alireza Kalaei

Alireza Kalaei is currently pursuing PhD studies at the University of Hong Kong. Prior to this, he completed his undergraduate studies at the Sharif University of Technology in 2020 and finished his MPhil studies at the University of Hong Kong in 2022. His awards include finishing his undergraduate degree as a first class honours graduate and receiving a prestigious type A PSG scholarship for the current PhD studies at the University of Hong Kong. His research interests include mechanical properties of materials, microstructural basis of strength, and computational materials science. On the topic of continuous dislocation dynamics, he has made important contributions to the development of numerical schemes and simulation codes.

IOP Publishing

Continuous Dislocation Dynamics Made Simple

Alfonso Ngan and Alireza Kalaei

Chapter 1

Introduction

1.1 Slip system and dislocations

The Institute of Materials, Minerals and Mining in the UK annually awards a distinguished materials scientist the prestigious Rosenhain Medal, named after Walter Rosenhain (1875–1934). One of Rosenhain's significant contributions to physical metallurgy was the discovery in ca. 1899 of slip traces on polished or etched surfaces of deformed metals, from which the concept of slip system was coined. As shown in figure 1.1(A), a slip system comprises a slip plane specified by a normal direction $\hat{n}$, on which plastic slip of the top half of the crystal occurs relative to the lower half by the vector $\vec{s}$. Rosenhain's observation of the slip traces on specimen surfaces after deformation confirmed that crystal plasticity takes place via discrete slip events on crystallographic slip systems.

However, the notion of slip systems alone could not explain the yield strength of metals, as the relative shear of two half crystals as rigid blocks on a specific slip system was expected to occur at a critical resolved shear stress (CRSS) on the order of $\mu/2\pi$ (μ being the shear modulus), which is more than two orders of magnitude higher than measured values of 10^{-4} to 10^{-3} μ. The concept of dislocations as the discrete quanta of plastic slip was first proposed by Egon Orowan, Michael Polanyi and Geoffrey I. Taylor in 1934, although the elastic stress–strain fields of dislocations as a possible type of crystalline defects were derived earlier by Vito Volterra in ca. 1907. In 1940, Rudolf E Peierls worked out the first theoretical prediction of the critical stress to move a dislocation, widely known in modern times as the Peierls stress, which falls in the correct order of magnitude of the measured yield strength. Direct observation by transmission electron microscopy in 1956 confirmed the existence of dislocations. As illustrated in figure 1.2, a glissile dislocation can be viewed as a line (not necessarily straight) lying on the slip plane to separate the latter into a slipped part from the unslipped part, while the slip amount $\vec{s}$ between the upper and lower half-crystals is now an atomic Burgers vector $\vec{b}$, which will restore the crystalline structure after the slip. The key defining parameters for a glissile

doi:10.1088/978-0-7503-6330-3ch1

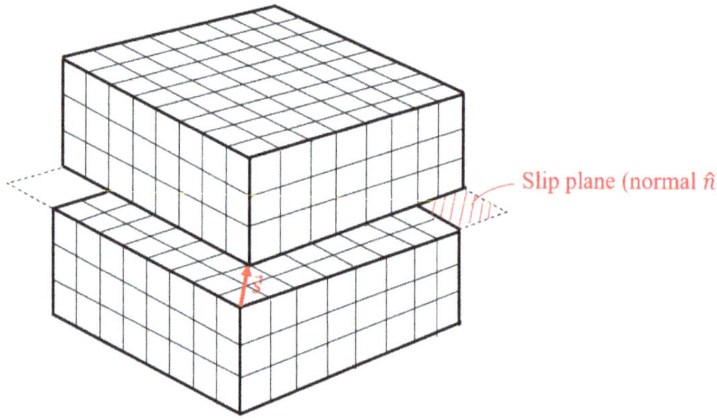

Figure 1.1. Plastic deformation of a crystal by slip on a slip system comprising a slip plane with normal $\hat{n}$, and a slip vector $\vec{s}$.

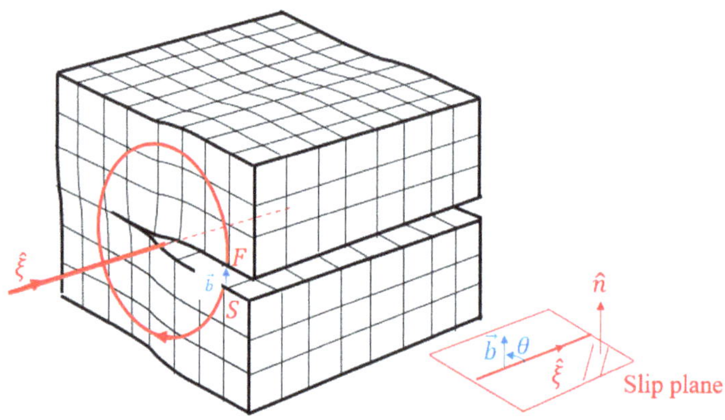

Figure 1.2. Glissile dislocation as line on the slip plane separating the slipped part of the slip plane from the unslipped part, where slip is by the atomic Burgers vector $\vec{b}$, which recovers the crystal periodicity after slip. Burgers vector $\vec{b}$ is defined as the vector $\overrightarrow{\mathrm{SF}}$ on the dislocated crystal, where S and F are the starting and ending points of a circuit broken by the dislocation drawn in the right-hand (RH) direction about the dislocation line direction $\hat{\xi}$. Therefore, for a glissile dislocation, the vectors $\vec{b}$ and $\hat{\xi}$ span the slip plane with normal $\hat{n}$, and the angle from $\hat{\xi}$ to $\vec{b}$ on the slip plane is the dislocation character θ.

dislocation segment include the dislocation line direction $\hat{\xi}$ and the Burgers vector $\vec{b}$, both of which must lie on the slip plane (i.e. $\hat{\xi} \cdot \hat{n} = \vec{b} \cdot \hat{n} = 0$), and a third useful parameter is the dislocation character θ, which is the inclination angle of $\vec{b}$ from $\hat{\xi}$ on the slip plane.

As the dislocation is viewed as a line feature, the line direction vector $\hat{\xi}$ can be defined to point along one of the two opposite directions along the dislocation. In the conventional 'RH/FS' rule, the sense of $\hat{\xi}$ is chosen first, and then a right-hand (RH) circuit about the positive $\hat{\xi}$ direction is drawn on the dislocated crystal as shown in figure 1.2. Due to the presence of the dislocation, the circuit is broken and then the vector $\overrightarrow{SF}$ connecting the starting point S to the finishing point F of the circuit is defined as the Burgers vector $\vec{b}$. Note that the $\vec{b}$ defined this way, known as the 'local' Burgers vector, would be a vector on the strained crystal with the presence of the dislocation, and so it is not quite exactly the slip vector in the perfect crystal that would restore the crystal structure after the dislocation has glided a long way away. The latter, known as the 'true' Burgers vector, is defined by a broken RH circuit drawn on the perfect crystal without the dislocation, which would map into a closed circuit after the dislocation is introduced (see figure 5.10(A) in chapter 5). The 'true' Burgers vector is defined as the vector $\overrightarrow{FS}$ joining the finishing point F to the starting point S of the circuit on the perfect crystal, hence the name FS/RH for the convention. On the other hand, unless otherwise stated, the 'local' Burgers vector defined in figure 1.2 is used throughout this book, and the difference with the 'true' Burgers vectors is very small since elastic straining of the crystal by the dislocation is very small. Finally, it is clear that the sense of the Burgers vector $\vec{b}$ depends on that of $\hat{\xi}$, i.e., the direction of $\vec{b}$ will be reversed if $\hat{\xi}$ is reversed for the same dislocation.

Starting from the work of Volterra, Orowan, Polanyi, Taylor and Peierls mentioned above, the physics of dislocations has been extensively studied and presented in many excellent texts. This book assumes the reader has basic knowledge of dislocation physics and mechanical properties, and references such as Hull and Bacon (2001), Smallman and Ngan (2014) and Hirth and Lothe (1992) in the reading list may be useful reminders. If needed, the reader may revise topics including:

- types of dislocations and their Volterra stress fields;
- elastic interaction forces between dislocations;
- Peach–Koehler force on dislocation due to resolved shear stress acting on the slip system;
- dislocation core and the Peierls–Nabarro model;
- dislocation generation by the Frank–Read source;
- dislocation–precipitate interactions by Orowan looping;
- Taylor's model for strain hardening.

1.2 Multiscale modeling of crystal plasticity and the meso-scale gap

Fast forwarding to modern times, the study of crystal plasticity, namely, the plastic deformation of crystals, has evolved into a subject relying heavily on computation, and in fact, a holy grail of computational materials science is the direct computation of the stress–strain behaviour of metals and alloys. Figure 1.3 summarizes the different types of computational tools for simulating crystal plasticity. At the bottom of the length/time scale, quantum mechanics methods based on density functional theory (DFT) and molecular dynamics (MD) based on known interatomic potentials can only handle at most a few hundred and a few million atoms, respectively, even with today's fastest computers. DFT and MD may therefore be good to study short dislocation segments or infinitely long dislocations with periodic boundary conditions applied, but not larger models. The next method up the scale ladder is discrete dislocation dynamics (DDD) which simulates discrete dislocation lines by computing the dynamics of the nodes joining the lines or discretized segments of the

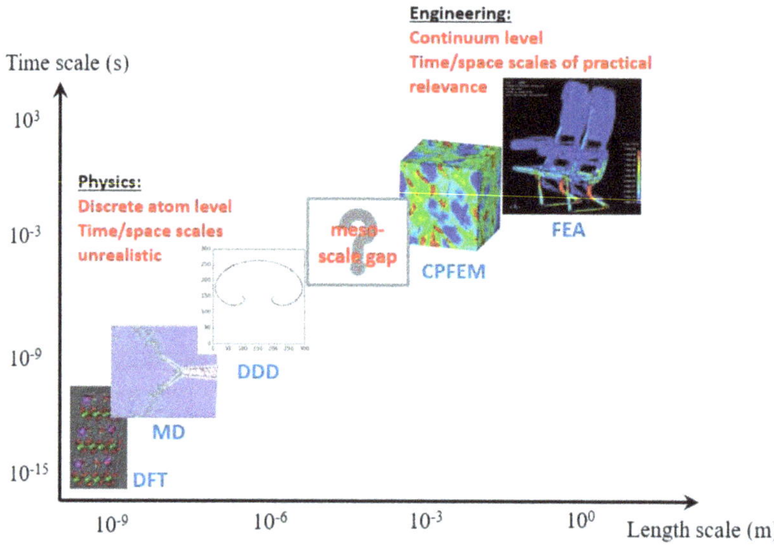

Figure 1.3. Modelling techniques for crystal plasticity at different length and time scales. (CPFEM image reprinted from Ghosh and Cheng (2018), copyright (2018), with permission from Springer.

lines. Currently, DDD can handle up to about a million nodes, using parallel computing with around a thousand processors for a few months of CPU time (Rao *et al* 2019). Assuming a nodal spacing of 0.1 μm, this gives a limit of ~0.1 m for the total dislocation length to be simulated, and with a specimen volume of 50 μm × 20 μm × 20 μm, the dislocation density simulated can be up to $\sim 10^{12}$ m m^{-3}. DDD is therefore good for simulating dislocations in micro-specimens but cannot deal with larger quantities of dislocations as in bulk samples or samples deformed to large plastic strains. The spatiotemporal scales involved in the 'physics' based methods of DFT, MD and DDD are therefore far off from those encountered in real engineering applications. To illustrate the problem, figure 1.4 shows typical dislocation microstructures in deformed metals. In figure 1.4(a) for pure Ni strained to just 2%, the dislocation distribution is already highly inhomogeneous over a length scale of a few microns, with dense patches at some places and rarefactions at others. In figure 1.4(b) for molybdenum deformed by indentation with a vibratory load, the dislocations form well-defined subgrains with sizes of around 2 μm. This type of non-uniform deformation microstructures, known as dislocation patterns, are the norm rather than the exception in deformed metals, and as shown in the examples in figures 1.4(a) and (b), the simulation size needed for studying them has to be at least a few microns which is already too large for DFT and MD, and the dislocation quantities involved would exceed what DDD can handle. The common phenomenon of dislocation patterning in fact occurs in the mesoscopic length scale in the range of a few to tens of microns, which is beyond the capability of the 'physics' based methods of DFT, MD and DDD in figure 1.3.

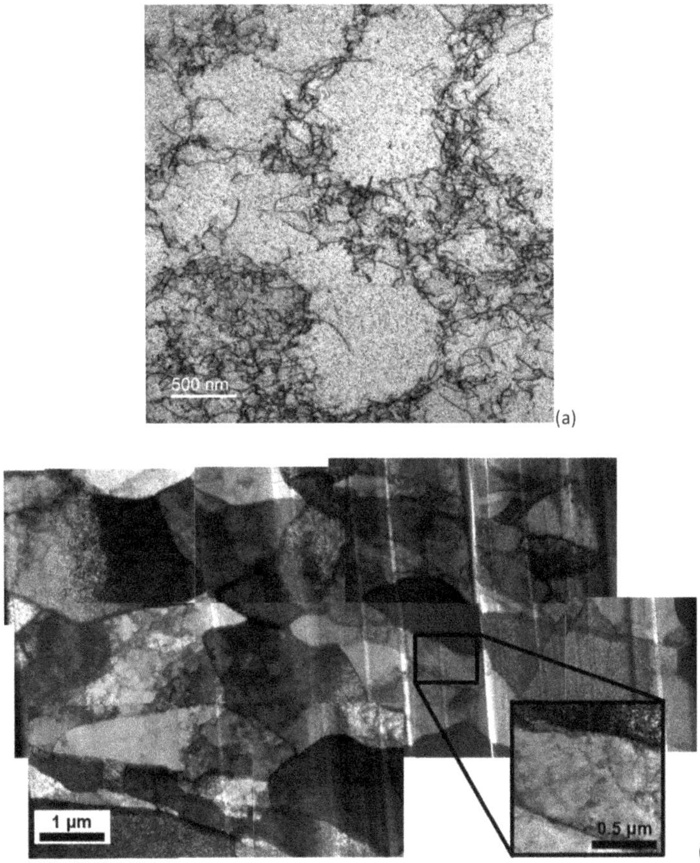

Figure 1.4. Transmission electron micrographs of deformed metals. (a) Pure Ni strained to 2%, showing groups of dense dislocations inhomogeneously distributed over an area smaller than 3 μm × 3 μm (image courtesy Dr Qingqing Sun). (b) Molybdenum deformed by vibratory indentation with displacement amplitude of 2 μm. In the view was originally one or two single grains in a specimen with initial grain size of 25 μm on average. After the vibratory indentation loading, dislocations form subgrains around 2 μm in size (reprinted from Siu and Ngan (2013), copyright (2013) with permission from Elsevier).

Jumping across the mesoscopic gap in the range of a few to tens of microns in figure 1.3, there is another group of simulation methods based on continuum mechanics. These methods take inputs from the lower-level physics-based models or measurements from experiments to predict the deformation of real-scale structures. At the top of the scale ladder, there is the well-known finite-element analysis (FEA) which aims to compute large structures from known constitutive laws of intrinsic material behaviour (e.g., relations of stress and strain, stress and strain rate, etc), which are usually measured experimentally. The next one down the ladder is the crystal-plasticity finite-element method (CPFEM), which is based on the concept of slip systems shown in figure 1.1. A key premise of CPFEM is that the plastic strain rate $\overline{\mathbf{L}}^{\mathrm{p}}$ (*aka* velocity gradient) of a deforming crystal is given by:

$$\overline{\mathbf{L}}^{\mathbf{p}} = \sum_{\alpha} \dot{\gamma}_{\alpha} \, \hat{s}_{\alpha} \otimes \hat{n}_{\alpha} \tag{1.1}$$

where α specifies the slip system, on which $\dot{\gamma}_{\alpha}$ is the shear strain rate, $\hat{s}_{\alpha}$ is the slip vector and $\hat{n}_{\alpha}$ the slip-plane normal. (The dyadic multiplication $\otimes$ between two vectors is explained in appendix A.) In addition, CPFEM assumes that $\dot{\gamma}_{\alpha}$ is given by 'flow rules' of the type (e.g., Dunne *et al* 2007):

$$\dot{\gamma}_{\alpha} = \begin{cases} \dot{\gamma}_0 \sinh\left[\beta\left(|\tau_{\alpha}| - \tau_{\alpha}^c\right)\right] \operatorname{sgn}(\tau_{\alpha}) & \text{for } |\tau_{\alpha}| > \tau_{\alpha}^c \\ 0 & \text{for } |\tau_{\alpha}| \leqslant \tau_{\alpha}^c \end{cases} \tag{1.2}$$

where $\dot{\gamma}_0$ and β are model material constants, τ_{α} is the resolved shear stress on the slip system and τ_{α}^c the CRSS, and $\operatorname{sgn}(\tau_{\alpha}) = 1$ if $\tau_{\alpha} > 0$ and -1 if $\tau_{\alpha} < 0$. Equation (1.2) essentially says that if the resolved shear stress τ_{α} is larger than the CRSS, the slip system will operate with a shear strain rate $\dot{\gamma}_{\alpha}$ that rises rapidly with the net difference between τ_{α} and CRSS, but if τ_{α} is smaller than the CRSS, then the slip system will not operate and then $\dot{\gamma}_{\alpha}$ will be zero. The strain rate *tensor* $\overline{\mathbf{L}}^{\mathbf{p}}$ (see appendix A) which sums up the contributions from all slip systems in the crystal (equation (1.1)) can then be added to the elastic deformation of the crystal under the applied loading to give rise to the elastoplastic deformation of the crystal (see section 5.6 in chapter 5 for details). CPFEM is therefore only a computational framework for the plastic strain rate and its integration with elastic deformation, and as such, any CPFEM scheme is as good as the flow rule used. However, the common flow rules proposed to-date such as that in equation (1.2) do not contain much physics of dislocations. The form in equation (1.2) only has the CRSS as the main physical parameter (thus bringing us back to the pre-dislocation, slip-system era of Rosenhain). There are other forms of the flow rule for use in CPFEM such as those involving strain hardening, but the common feature is still that detailed dislocation physics are not incorporated. One exception is the 'discrete-continuum model' (DCM) that employs DDD as the flow rule for computing the strain rate $\dot{\gamma}_{\alpha}$ for use in a CPFEM framework (Cui *et al* 2022, see also section 7.1.1 in chapter 7), but as discussed above, the amount of dislocations handleable by DDD is rather limited.

Having understood the 'physics' and 'engineering' based methods in figure 1.3, the mesoscopic gap between the two groups represents not only a discontinuity in the spatiotemporal scale of the simulation methods, but worse still, the abrupt abandon of the rich body of knowledge on dislocation physics gathered since Volterra's time. Thus, even though the 'engineering' based methods are often used to simulate large systems in real applications, such simulations lose contact with our rich knowledge on dislocations physics that govern microstructures such as those in figure 1.4.

1.3 CDD and why this book?

This book advocates the use of *dislocation density* as the basic variable for developing simulation tools to bridge the mesoscopic gap in figure 1.3. What is dislocation density and why is this preferable? As will be reviewed in detail in chapter 2, dislocation density is a continuum (or coarse-grained) measure of the quantity (i.e., total line length) of dislocations residing in a unit material volume. Recalling that in the DDD method at one end of the mesoscopic gap, the simulation variables are the positions of the dislocation nodes or segments, therefore the computation quickly gets saturated as the dislocation quantity becomes large. However, a simulation strategy based on density, known commonly as continuous dislocation dynamics (CDD), would not get saturated this way, since a high quantity of dislocations is represented by just a large value of the density variable. Figure 1.5 shows an example of CDD simulation of dislocation density distribution developed into a subgranular structure under vibratory load, capturing the essence of the experimental observation in figure 1.4(b). In contrast with CPFEM at the other end of the mesoscopic gap, a CDD simulation strategy can only be built on detailed considerations of dislocation dynamics, thus CDD can be safely placed in the category of physics-based methods in figure 1.3, yet the simulation size can overlap with that of CPFEM. These are the key reasons why CDD methods are promising to bridge the mesoscopic gap between DDD and CPFEM in figure 1.3.

Using density to describe dislocations in fact dated back to the work of Peierls on lattice friction for dislocation motion. The stress and strain fields of dislocations as derived by Volterra are based on linear elasticity, and as such, they exhibit a $1/r$ type of singularity where r is distance from the dislocation centre. As stress and strain cannot be infinite, Volterra's solutions are for a fixed displacement vector $\vec{b}$ for the slipped part of the slip plane (figure 1.2) up to some cutoff distance near the dislocation centre, and then the material inside the cutoff, which would otherwise be subjected to infinite stress and strain is ignored. Peierls' subsequent work was to add

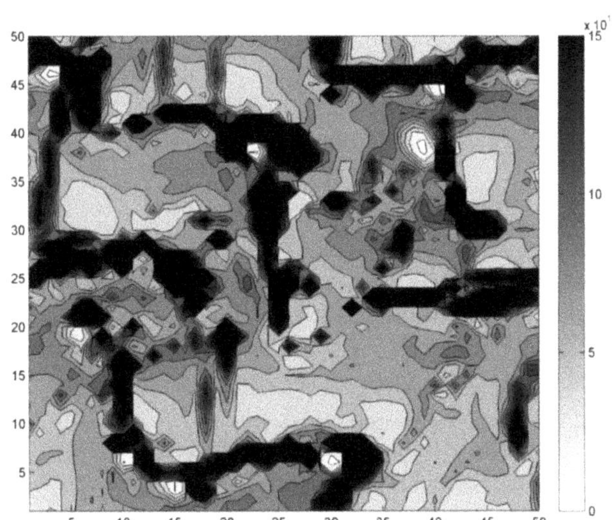

Figure 1.5. CDD simulation of dislocation density distribution developed into a cellular structure under a vibratory load, akin to figure 1.4(b). Grey scale indicates dislocation density in units of m^{-2}. Cheng *et al* (2015), reprinted by permission of the publisher (Taylor & Francis Ltd, http://www.tandfonline.com).

back the core of the dislocation but instead of using a fixed Burgers vector up to the very centre, he prescribed the shear displacement in the core by a varying 'disregistry' function $u(r)$, the derivative of which, $\rho = du/dr$, can be interpreted as a density distribution of dislocations with infinitesimal Burgers vectors (see section 4.3, chapter 4 for details). In ca. 1953, J F Nye introduced a tensor, known as the Nye tensor, to represent the density of geometrically necessary dislocations (GNDs). The exact evolution law for the Nye tensor is known and is reviewed in section 3.1 of chapter 3. For coarse-grained density description of many dislocation lines, there are also a number of phenomenological models, such as the Walgraaf–Aifantis type models based on the assumption that dislocation motion is diffusive (see exercise 3.4, chapter 3), or the Kocks–Mecking type models involving dislocation storage and annihilation:

$$\frac{\mathrm{d}\rho}{\mathrm{d}\gamma} = A\sqrt{\rho} - B\rho \tag{1.3}$$

where ρ is dislocation density, γ is plastic strain, and A and B are model constants (Kocks 1976). The density evolution laws in these ad hoc models are not exact since they do not obey the requirement that dislocations are connected lines while they move; for this reason, these models will not be dealt with in this book.

At this juncture, an important remark for readers who are already familiar with the CDD literature. Since only the motion of geometrically necessary dislocations (GNDs, reviewed in section 2.3.1 of chapter 2) will produce net plastic deformation of the crystal while the remaining 'dipole' contents of statistically stored dislocations (SSDs) will have their shape-change effects cancelled out on average, it may therefore be a common belief that all CDD theories should focus only on the GND density ρ_{GND}, and the SSD density ρ_{SSD} can be ignored. Unfortunately, this belief is incorrect for *coarse-grained* simulations, in which the pixel is not small so that it may contain some SSD contents. To understand this, consider the short-range interactions of dislocations (discussed later in section 5.5, chapter 5) which give rise to the well-known Taylor hardening that depends on the square root of the total dislocation density:

$$\tau_{\text{Taylor hardening}} = \mu b\sqrt{\rho_{\text{total}}} \tag{1.4}$$

Here, b is the magnitude of the Burgers vector, and $\rho_{\text{total}} = \rho_{\mathrm{GND}} + \rho_{\mathrm{SSD}}$ is the total dislocation density. Therefore, the presence of SSDs will increase the resistance of dislocation motion and hence affect the dynamics and evolution of the GNDs. Thus, in *coarse-grained* simulations in which SSD contents are present, the evolution of ρ_{GND} cannot be separated from that of ρ_{SSD}, and this is why any CDD theory that considers the GND density alone would be inadequate (see chapter 7 for a comprehensive review of the CDD literature and a more detailed discussion on this point).

Since the evolutions of ρ_{GND} and ρ_{GND} are coupled in a coarse-grained sense, this book focuses on a special definition of the dislocation density known as the 'all-dislocation density' (ADD), which includes both GND and SSD contents, and in fact, avoids their somewhat arbitrary distinction that some readers may not be fond of. The prime objective of this book is therefore to introduce a special version of CDD based on the ADD, for which the kinematics law is miraculously known exactly, also in a rather simple form!

Chapter 2 formally introduces the dislocation density and the various versions of it. The kinematics and dynamics of discrete dislocation lines represented as density distributions (in the sense of the Peierls–Nabarro model) are derived in chapters 3 and 4. Then, in chapters 5 and 6, the kinematics, dynamics and numerical simulation of *coarse-grained* dislocation microstructures (in which individual dislocation lines are smeared out into a coarse-grained density) are discussed. The book uses mathematics and standard solid mechanics representations understandable by a graduate student, but if a quick reminder is needed, the reader is suggested to first go through appendix A or refer to it when needed. Simulation codes are given and explained throughout the book, and these are written in MATLAB—hence a basic knowledge of this language will be useful. Other than these and the fundamental knowledge of dislocation physics mentioned in section 1.1, no further prerequisite is required, and the various concepts needed will be explained as one reads along.

References

Cheng B, Leung H S and Ngan A H W 2015 Strength of metals under vibrations—dislocation-density-function dynamics simulations *Phil. Mag.* **95** 1845–65

Cui Y, Wang T, Luo S, Li Z and Li Z 2022 A discrete-continuum model of three-dimensional dislocation elastodynamics *Int. J. Plasticity* **152** 103221

Dunne F P E, Rugg D and Walker A 2007 Lengthscale-dependent, elastically anisotropic, physically-based hcp crystal plasticity: application to cold-dwell fatigue in Ti alloys *Int. J. Plasticity* **23** 1061–83

Ghosh S and Cheng J 2018 Adaptive multi-time-domain subcycling for crystal plasticity FE modeling of discrete twin evolution *Comput. Mech.* **61** 33–54

Hirth J P and Lothe J 1992 *Theory of Dislocations* 2nd edn (Malabar, FLlorida: Krieger Publishing)

Hull D and Bacon D J 2001 *Introduction to Dislocations* 4th edn (Oxford: Butterworth-Heinemann)

Kocks U F 1976 Laws for work-hardening and low-temperature creep *J. Eng. Mater. Technol.* **98** 76–85

Rao S I, Woodward C, Akdim B, Antillon E, Parthasarathy T A, El-Awady J A and Large-scale D M D 2019 dislocation dynamics simulations of strain hardening of Ni microcrystals under tensile loading *Acta Mater.* **164** 171–83

Siu K W and Ngan A H W 2013 Oscillation-induced softening in copper and molybdenum from nano-to micro-length scales *Mater. Sci. Eng.* A **572** 56–64

Smallman R E and Ngan A H W 2014 *Modern Physical Metallurgy* (Amsterdam: Elsevier)

IOP Publishing

Continuous Dislocation Dynamics Made Simple

Alfonso Ngan and Alireza Kalaei

Chapter 2

What are dislocation densities?

2.1 Defects and their densities

Before discussing dislocations and their densities, let us take a step back to look at material microstructures in general, of which dislocations are just one type. In a first course on materials science, some instructors or textbooks will use the four basic geometric forms of point, line, plane and volume to classify microstructures. Thus, atomic microstructures like solutes or vacancies are point defects, dislocations or disclinations are line defects, grain boundaries, cracks and other interfaces are planes or surfaces, and second-phase precipitates and voids are volumes (table 2.1).

In the same elementary course or textbook, the second topic to be introduced may be diffusion of point defects. There, a concentration is introduced as follows:

$$C(\vec{r}) = \frac{\Delta n}{\Delta V} \tag{2.1}$$

where ΔV is a small volume voxel around point $\vec{r}$, and Δn is the number of point defects inside ΔV. This is followed by Fick's two laws for diffusion, the second one of which is:

$$\frac{\partial C}{\partial t} = -\vec{\nabla} \cdot \vec{J} \tag{2.2}$$

where $\vec{\nabla} = \hat{e}_x(\partial/\partial x) + \hat{e}_y(\partial/\partial y) + \hat{e}_z(\partial/\partial z)$ is the gradient or 'del' operator, and $\vec{J}$ is diffusion flux.

So much so for the reminder of our first course on materials science, but things so far prepare us well to tackle dislocation densities based on mathematical generalization. In particular, through equations (2.1) and (2.2) for point defects, we can appreciate a few important concepts which will be encountered frequently in the rest of this book, namely, *density, coarse graining, kinematics law and dynamics closure*:

doi:10.1088/978-0-7503-6330-3ch2

Table 2.1. Objects and their densities in space.

Objects	Microstructural examples	Density definition	Measurer	How to measure
Points	Vacancies, solutes	Number per unit volume (#/m^3)	A small volume	Using a small volume to count how many point defects are present inside it
Lines	Dislocations, disclinations	Number threading unit area (#/m^2)	A small area	Using a small area to count how many lines are threading it
Planes	Grain boundaries, interfaces	Number intersecting unit line (#/m)	A short line segment	Using a short line to count how many planes are intersecting
Volumes	Second phase, voids	Volume fraction (v/v)	A random point	Moving a point around a volume region to sense probability of the point residing in the second phase or void

Coarse-grained density—The C defined in equation (2.1) is a density which, when used in conjunction with its governing equation (2.2), is a scalar function of space and time, i.e., $C = C(\vec{r}, t)$. Through C, the point defects are no longer viewed as individual vacancies or atomic solutes located in a crystal, but as a coarse-grained microstructural feature of the crystal defined by the density C, which is now a smeared-out representation of the local quantity of the point defects (figure 2.1). The diffusion problem is therefore transformed from one that tracks the movement of each vacancy or solute in the crystal, to one that aims to solve the differential equation in (2.2) for $C(\vec{r}, t)$. The coarsening of the resolution scale of the problem from atomic to the typical spatiotemporal scale over which C varies is known as *coarse graining*. Alternatively, we may also say that the problem is transformed from a *discrete* consideration of the individual point defects, to a *continuum* consideration of the smooth density function C over space and time domains. The advantage of the continuum formalism is obvious—this allows any quantity of point defects to be modelled, since a small quantity is represented by a low value of C while a big quantity is represented by a large C. On the contrary, the discrete formalism will be quickly handicapped for any realistic quantity of point defects; say 1 kg of a crystal will easily contain many moles of point defects that cannot be handled by atomic models like molecular dynamics even with today's fastest computers. Back to the elementary course, the reader will remember the use of Fick's second law (equation (2.2)) to solve problems such as case-hardening of steels, which will be intractable with the discrete formalism. If the reader does not object to the

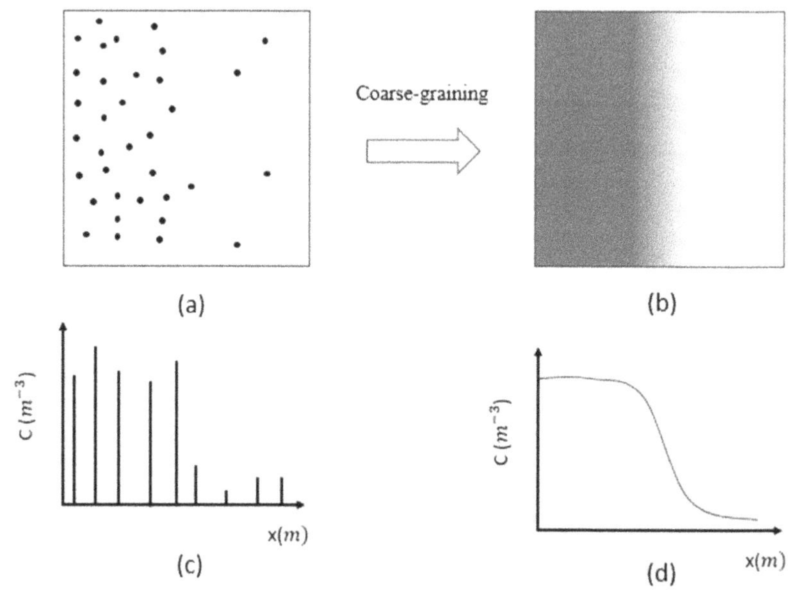

Figure 2.1. Illustrating coarse-graining for point defects. (a) Points spatially distributed. (b) Coarse-grained density distribution. (c) Counts of point density along a spatial direction. (d) Smeared out density along the spatial direction.

continuum formalism for point defects, they are well set for appreciating the same concept for dislocations, which is the subject matter of this book.

Kinematics law and dynamics closure—equation (2.2) is an evolution equation, or *kinematics law* for the defect density C. It is so because this equation governs the function C, which if solved will reveal how the defect density evolves over the space and time domains. However, equation (2.2), which is Fick's second law, is not sufficient for solving C, and the reader will remember Fick's first law, which may be cast as follows:

$$\vec{J} = -\frac{D}{kT} C \, \vec{\nabla} \, \mu \tag{2.3}$$

Here μ is chemical potential of the point defects which may be expressed as

$$\mu = p\Omega + kT \ln C \tag{2.4}$$

where p is hydrostatic stress and Ω is atomic volume of the point defect, and kT is the thermal energy. Substituting equations (2.3) and (2.4) into (2.2) yields:

$$\frac{\partial C}{\partial t} = D \, \nabla^2 \, C + \frac{D\Omega}{kT}(\vec{\nabla} C \cdot \vec{\nabla} \, p + C \, \nabla^2 \, p) \tag{2.5}$$

which is now a *closed* governing equation for C on its own, provided that $p(\vec{r})$ and the initial and boundary conditions for C are known. Introducing equations like (2.3) and (2.4) is known as *dynamics closure*, which is a strategy to close the problem by adding force effects to the kinematics law, which is only for the motion of the system. As will be discussed in later chapters, the correct kinematics law and its dynamics closure scheme(s) need to be established for dislocations, after properly defining their coarse-grained densities.

Density measurer—Having seen the coarse-grained density of point defects defined in equation (2.1), how should we define the density of other types of defects in table 2.1? For point defects, the 'measurer' for the defect quantity is a small material volume δV, so that the unit for C is #/m^3 (# means number). The reader may also remember, perhaps also from the elementary course, that in metallography, the grain size is measured by drawing a straight line across many grains in an optical or electron micrograph and counting the number of grain boundaries intersecting the line. Hence, for surface defects in general, their density, again in the coarse-grained sense, may be defined as

$$\vec{S}(\vec{r}) = \frac{\Delta n}{\Delta L} \Delta \hat{L} \tag{2.6}$$

where ΔL is a short line segment along direction $\Delta \hat{L}$ at around point $\vec{r}$, and Δn is the number of surface defects intersecting ΔL (figure 2.2). In this case, the density measurer is a short line length intersecting the surface defects, and the unit of $\vec{S}$ is #/m. For volume defects, their density is just the volume fraction f_V, which is dimensionless. We may use a point as the measurer for the f_V of a second phase, i.e., by randomly positioning the point over the specimen, the probability of the point falling inside the second phase will be the f_V:

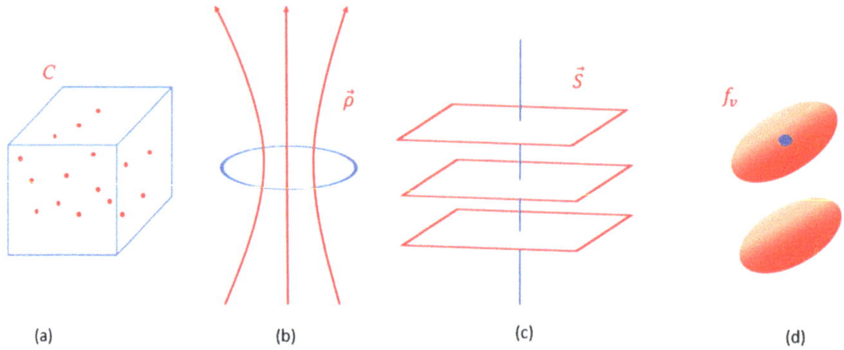

Figure 2.2. Showing density measurers (in blue) for (a) point, (b) line, (c) surface and (d) volume defects (in red).

$$f_v = \frac{\text{volume of second phase}}{\text{total volume of sample}} \\ = \text{probability of a random point falling into the second phase} \tag{2.7}$$

Therefore, going from point, surface, to volume defects, the density measurer goes from a volume, a line, to a dimensionless point. Then, what should be the density measurer for line defects? The obvious answer must be a small area (see table 2.1). Therefore, the density of line defects must be of the form:

$$\vec{\rho}(\vec{r}) = \frac{\Delta n}{\Delta A}\hat{\rho} \tag{2.8}$$

where $\hat{\rho}$ is the unit direction of the lines, ΔA is a small area normal to $\hat{\rho}$ around point $\vec{r}$, and Δn is the number of line defects threading ΔA. $\vec{\rho}$ has the unit #/m^2. Looking at table 2.1 and figure 2.2, it is interesting to see that the dimensions of the defects and the density measurer always add up to 3, the dimension of space. This is another nice example of the beauty of mathematics!.

2.2 Fine grained dislocation density

2.2.1 Bundle of parallel dislocations

The line density defined in equation (2.8) readily applies to a bundle of parallel dislocation lines (figure 2.3), as first advocated by E Kröner in the late 1950s. For a bundle of parallel dislocations, the density magnitude $\rho = \lceil \vec{\rho} \rceil$ also happens to be the line length per unit volume:

$$\rho(\vec{r}) = \frac{\Delta n}{\Delta A} = \frac{\Delta L}{\Delta V} \tag{2.9}$$

where ΔV is a small volume around point $\vec{r}$, and ΔL is the total length of dislocation lines inside ΔV. Therefore, the unit of ρ may also be expressed as m m^{-3}.

The bundle of parallel dislocations in figure 2.3 are not merely lines because they also possess the Burgers vector $\vec{b}$. Therefore, the dislocation bundle should be

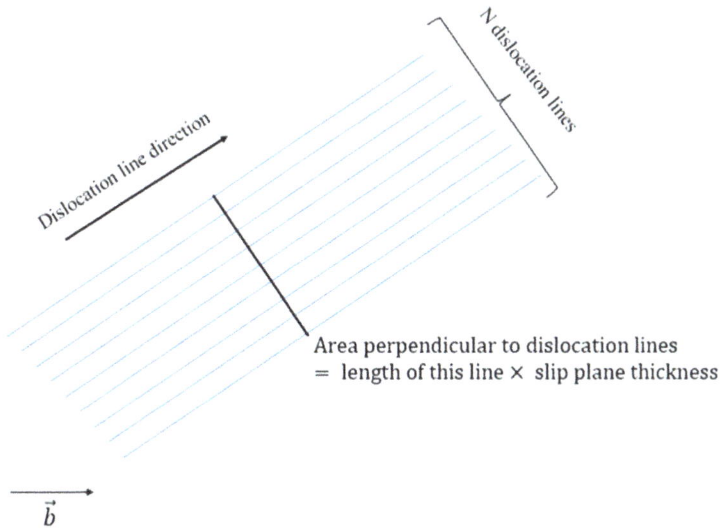

Figure 2.3. A bundle of parallel dislocations of the same Burgers vector.

specified by not just the density vector $\vec{\rho}$, but also the $\vec{b}$. In mathematical mechanics, two vectors in 3D space can be jointly represented by a tensor which is really a 3 by 3 matrix, and for the parallel dislocation bundle, J F Nye proposed the following tensor

$$\boldsymbol{\alpha} = \vec{b} \otimes \vec{\rho} \tag{2.10A}$$

which is just the 3 × 3 matrix

$$\{\alpha_{ij}\} = \begin{pmatrix} \alpha_{11} & \alpha_{12} & \alpha_{13} \\ \alpha_{21} & \alpha_{22} & \alpha_{23} \\ \alpha_{31} & \alpha_{32} & \alpha_{33} \end{pmatrix} = \begin{pmatrix} b_1\rho_1 & b_1\rho_2 & b_1\rho_3 \\ b_2\rho_1 & b_2\rho_2 & b_2\rho_3 \\ b_3\rho_1 & b_3\rho_2 & b_3\rho_3 \end{pmatrix}$$

where i is row index and j column index, and the matrix elements are:

$$\alpha_{ij} = b_i \rho_j \tag{2.10B}$$

Here, b_i and ρ_j (i, j = 1–3) are the three Cartesian components of $\vec{b}$ and $\vec{\rho}$, respectively. Physically, equation (2.10B) says that α_{ij} is the total magnitude of the Burgers vector component along direction i threading the unit plane normal to j.

2.2.2 Sub-dislocation density—intensive dislocation microstructures

While the line density $\vec{\rho}$ defined in equation (2.8) and the Nye tensor $\boldsymbol{\alpha}$ defined in equation (2.10) are readily applicable to bundles of parallel dislocations as illustrated in figure 2.3, real dislocation microstructures seldom develop such configurations. Whereas realistic dislocation microstructures will be considered in

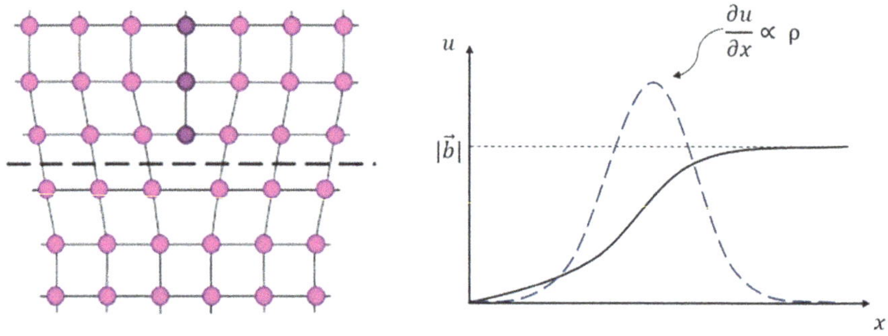

Figure 2.4. Density distribution inside a single dislocation and displacement disregistry function u.

the next section, here we note another scenario where $\vec{\rho}$, α and their kinematics laws are applicable. This concerns the dislocation contents (density) inside the core of a single dislocation, in the context of the classical Peierls–Nabarro model (figure 2.4). In this model, the crystal is cut into two halves along the slip plane and are displaced relative to one another via a disregistry function $u(x)$, where x is the coordinate perpendicular to the dislocation line on the slip plane. The boundary conditions for u are: $u(-\infty) = 0$ and $u(+\infty) = |\vec{b}\,| = b$, and the dislocation centre is roughly at $x = 0$. In this context, the dislocation density on the slip plane is given as:

$$\vec{\rho}_{P-N}(x) = \frac{1}{bd}\frac{du(x)}{dx}\hat{\xi} \tag{2.11}$$

where $\hat{\xi}$ is the unit direction of the dislocation, and d is the slip-plane thickness.

When modelling sub-dislocation contents, the graining level of the density has to be very fine ($\ll$ core width of the dislocation), and the corresponding density profiles are known as *intensive microstructures* because of the fine graining involved.

2.3 Coarse grained dislocation density—extensive dislocation microstructures

As stated in chapter 1, the purpose of continuous dislocation dynamics (CDD) is a mesoscopic crystal plasticity methodology capable of handling large quantities of dislocations well beyond what MD and DDD can handle. The dislocation microstructures concerned would typically have feature sizes on the order of a micron, and the system size can be as large as hundreds of microns. For such microstructures, which are referred to as *extensive* hereafter, the objective is not to resolve individual dislocations or even bundles of parallel dislocations, but to work with a resolution scale coarse enough for the dislocation contents inside to be treated as fairly uniform, while still fine enough to capture the coarser scaled fluctuations of interest such as subcells, persistent slip bands or pileups at obstacles. In such a coarse-grained context, the parallel bundle and sub-dislocation scenarios for which the $\vec{\rho}$ and α are applicable would be rather irrelevant. We therefore need to reconsider the notion of dislocation density for the coarse-grained context for modelling extensive microstructures.

2.3.1 Geometrically necessary dislocations (GNDs) and statistically stored dislocations (SSDs)

The dislocation density of a metal can typically rise to $\sim 10^{16}$ m^{-2} after heavy deformation, so that there will be $\sim 10^2$ dislocations threading a resolution pixel at a graining level of 0.1 μm. With so many dislocations in the graining pixel, they likely will have different line orientations even if they belong to the same slip system (figure 2.5(a)). Traditionally, the total dislocation density ρ_{total} is considered to comprise two components of (i) geometrically necessary dislocation (GND) density ρ_{G}, and (ii) statistically stored dislocation (SSD) density ρ_{S}, i.e.,

$$\rho_{\text{total}}(\vec{r}) = \rho_{\text{G}}(\vec{r}) + \rho_{\text{S}}(\vec{r}) \tag{2.12}$$

The ρ_{total} here is in the sense of equation (2.9), i.e., $\rho_{\text{total}} = \Delta L/\Delta V$ where ΔL is the total length of all dislocations within the volume pixel ΔV irrespective of their orientation. The GND density is the vector sum of the dislocation contents, i.e.,

$$\vec{\rho}_{\text{G}}(\vec{r}) = \sum_{i \in \Delta V} \vec{\rho}_i(\vec{r}) \tag{2.13}$$

where $\vec{\rho}_i$ is density of the ith parallel dislocation bundle that falls within ΔV, and $\rho_{\text{G}} = |\vec{\rho}_{\text{G}}|$. The SSD density ρ_S is the net of ρ_{total} after deducting ρ_G, and physically it represents the dipole contents in the ρ_{total}. Figure 2.5 illustrates the GNDs and SSDs.

2.3.2 All-dislocation density (ADD)

The GNDs and SSDs only represent, respectively, the net vector-sum and dipole contents of the dislocation population under consideration. As will become clear in chapter 5, this representation is not satisfactory, since no exact kinematics law is known for the GNDs or the SSDs. Figure 2.6 illustrates another coarse-grained representation that preserves all the dislocation bundles of different line directions within ΔV, instead of summing them into the $\vec{\rho}_{\text{G}}$ as in equation (2.13) and hence losing the SSD contents. In this representation, the 'all-dislocation' density

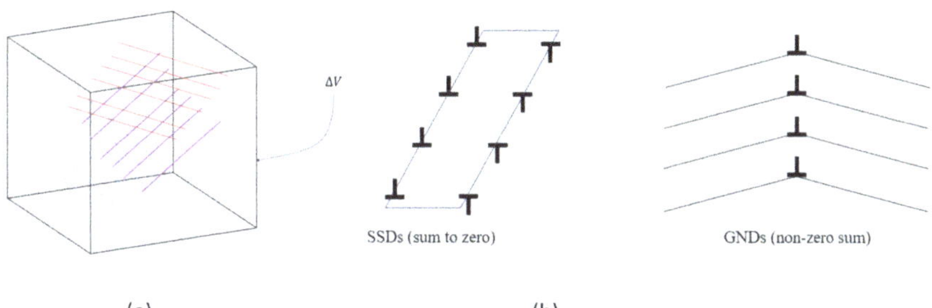

Figure 2.5. (a) Coarse-grained pixel ΔV (size $\sim$0.1 μ) containing dislocations of different directions. (b) The reduced representation by SSDs and GNDs.

Figure 2.6. Definition of all-dislocation density (ADD) $\varrho(\vec{r}, \theta)$. (a) An example of dislocation microstructure (a cellular pattern of dislocations in pure Ni strained to 2%; image courtesy Dr Qingqing Sun). (b) A pixel around point $\vec{r}$ in the space domain, with schematic dislocation lines of different orientations inside. (c) Polar plot of the $\varrho(\theta)$ dependence for the pixel at $\vec{r}$, where θ is dislocation character (distance of a point on contour at θ from origin $= \varrho(\theta)$).

(ADD) $\varrho(\vec{r}, \theta)$ is defined over both the spatial and dislocation-character domains as (Ngan 2017):

$$\varrho(\vec{r}, \theta)\, d^3r\, d\theta = \text{line length of dislocations of character} \in [\theta, \theta + d\theta] \text{ within volume } d^3r \text{ at } \vec{r} \tag{2.14}$$

As figure 2.6 shows, the dislocation contents inside a pixel at $\vec{r}$ (figure 2.6(b)) are collected by their dislocation character θ into the function $\varrho(\theta)$, which is schematically shown as a polar plot in figure 2.6(c). Such a representation preserves all dislocations of different character θ, at different positions $\vec{r}$ in the spatial domain. Not only is the full information preserved, the exact kinematics law for $\varrho(\vec{r}, \theta)$ is also available, as will be shown in chapter 5.

2.4 The three distinctive length scales for dislocation plasticity

Finally, let us introduce three length scales for pursuing dislocation-plasticity simulation that are distinctive (figure 2.7). The intensive and extensive length scales have already been explained above: intensive refers to sub-dislocation details where the pixel size has to be smaller than the dislocation core width, whereas extensive refers to a coarse-grained representation in which the resolution pixel has to be larger than the mean spacing between individual dislocations but smaller than the typical length scale over which the dislocation microstructure of interest fluctuates. In between these two, there is also the discrete length scale, where dislocations are treated as discrete lines without consideration of their core details. The discrete scale is the premise for discrete dislocation dynamics (DDD), while the intensive and extensive scales are the realm of CDD. When pursuing dislocation plasticity

Three length scales for dislocation plasticity	Intensive	Discrete	Extensive
	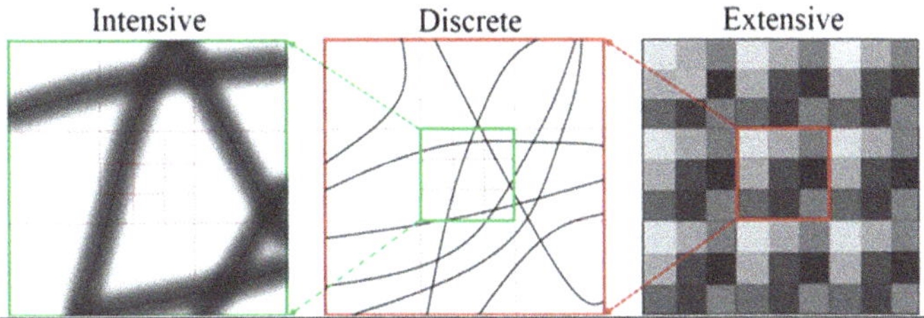		
Definition	Burgers vector content within the dislocation core is treated as a continuous distribution of density, as in the Peierls-Nabarro model	Dislocations treated as individual lines without consideration of cores	Highly coarse-grained, not resolving individual dislocations which are represented as a smeared-out density
Pixel size	core width >> pixel >> atomic	pixel >> dislocation core width	size of dislocation "patterns" >> pixel >> mean spacing between dislocations
Interactions	Elastic + specific treatment for short-range interactions	Elastic + specific treatment for short-range interactions	Elastic+ short-range interactions via Taylor hardening ($\propto \sqrt{\rho}$)
Peierls stress	Treated in relation to lattice disregistry (γ-surface)	Assumed constant	Assumed constant

Figure 2.7. Distinctive length scales for dislocation plasticity.

modeling, the choice of the length scale has to be determined at the outset, since the dynamics closure considerations of the three length scales in figure 2.7 are very different. For the intensive scale, core interactions and lattice resistance to dislocation motion (the Peierls stress) need to be explicitly considered. The lattice resistance may be assumed to be a constant friction for discrete and extensive scales, but considerations for mutual dislocation interactions are also quite different between these two length scales. In short, the modelling length scale has to be determined at the onset, as there is no 'one-size-fitting-all' scheme for dynamics closure for the different length scales. The different considerations for dynamics closure for the intensive and extensive length scales will be discussed in later chapters.

References

Kröner E 1958 *Kontinuumstheorie der Versetzungen und Eigenspannungen* (Berlin: Springer)

Ngan A H W 2017 Dislocation-density kinematics: a simple evolution equation for dislocation density involving movement and tilting of dislocations *MRS Commun.* **7** 583–90

Nye J F 1953 Some geometrical relations in dislocated crystals *Acta Metall* **1** 153

IOP Publishing

Continuous Dislocation Dynamics Made Simple

Alfonso Ngan and Alireza Kalaei

Chapter 3

Kinematics of intensive line density: evolution laws and simulation codes

For the line density $\vec{\rho}$ defined in equation (2.8) in chapter 2, its exact kinematics law exists, and in fact, in different equivalent forms. In this chapter, we will introduce these laws, as well as our first MATLAB programme for implementing such a law to simulate the motion of lines.

3.1 Maxwell's kinematics law for line density

To recap, the line density $\vec{\rho}$ defined in equation (2.8) is the number of lines threading unit area and a vector along the lines concerned. Figure 3.1 shows a measurer surface ΔA, bounded by curve C, used to count lines of density $\vec{\rho}$ moving with local velocity $\vec{v}$ in the 3D space around ΔA. During a short time interval δt, lines within a small cross-sectional area

$$(v\ \delta t) \times [(\hat{v} \times \hat{\rho}) \cdot d\vec{r}] = -(\hat{\rho} \times \vec{v}) \cdot d\vec{r}\ \delta t$$

will cut across a small element $d\vec{r}$ of curve C, where v is the magnitude and $\hat{v}$ the unit direction of $\vec{v}$. Therefore, during δt, the number of lines cutting through $d\vec{r}$ will be $|\vec{\rho}|$ times the above area, and hence,

$$\text{lines cutting through } d\vec{r} \text{ of } C \text{ per unit time} = -(\vec{\rho}\times\vec{v}) \cdot d\vec{r} \tag{3.1}$$

Integrating equation (3.1) over contour C then gives the net number of lines moving into surface ΔA per unit time as:

$$-\oint_C (\vec{\rho} \times \vec{v}) \cdot d\vec{r} = -\iint_{\Delta A} \vec{\nabla} \times (\vec{\rho} \times \vec{v}) \cdot \hat{n}\ dA \tag{3.2}$$

where Stoke's theorem is used with $\hat{n}$ being the unit normal of infinitesimal dA within the measurer area ΔA. On the other hand, the accumulation rate of lines threading ΔA may also be written as

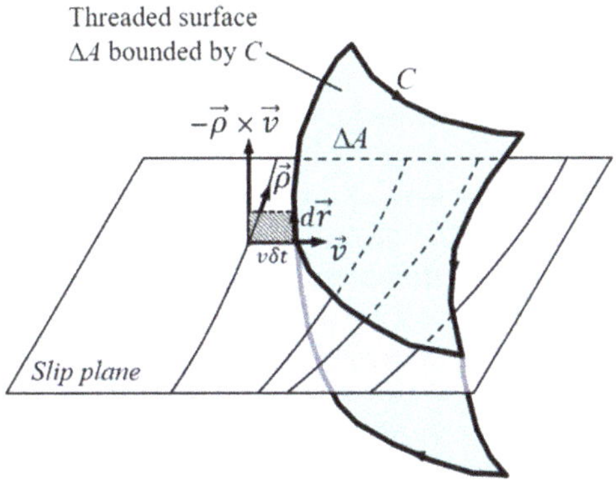

Figure 3.1. Lines moving in space counted by a surface element ΔA.

$$\frac{d}{dt}\iint_{\Delta A} \vec{\rho} \cdot \hat{n}\, dA = \iint_{\Delta A} \dot{\vec{\rho}} \cdot \hat{n}\, dA \tag{3.3}$$

where the dot over $\vec{\rho}$ means its time derivative. Hence, equating (3.2) and (3.3) and noting that the size of ΔA is arbitrary, we have

$$\dot{\vec{\rho}} = -\vec{\nabla} \times (\vec{\rho} \times \vec{v}) \tag{3.4}$$

which is the kinematics law for $\vec{\rho}$ that we attempt to seek. Equation (3.4) is actually compatible with Maxwell's equation for electromagnetism $\dot{\vec{B}} = \vec{\nabla} \times \vec{E}$ where $\vec{B}$ is magnetic field and $\vec{E}$ is the electric field induced by a change in $\vec{B}$ with time. But from Lorentz' law, as the magnetic flux $\vec{B}$ moves with velocity $\vec{v}$ relative to a unit test charge, the force acting on the test charge is $\vec{E} = -\vec{B} \times \vec{v}$, which is just the same induced electric field in Maxwell's equation. Hence, combining Maxwell's and Lorentz' equations, the magnetic flux $\vec{B}$ moves according to $\dot{\vec{B}} = -\vec{\nabla} \times (\vec{B} \times \vec{v})$, which is the same as equation (3.4). For this reason, equation (3.4) is referred to in the rest of this book as Maxwell's equation as an evolution law for moving lines in space. Note also that equation (3.4) as the kinematics law for line density is analogous to Fick's second law (equation (2.2)), which is the kinematics law for point density, and both are not in closed form. We have seen that for point density, equations (2.3) and (2.4) are needed to close equation (2.2); similarly, for line density, some dynamics law will be needed to prescribe the velocity field $\vec{v}(\vec{r})$ ($\vec{r}$ being a general field point) in order to close equation (3.4), and this will be considered in chapter 4.

For the Nye tensor $\boldsymbol{\alpha}$ defined in equation (2.10), its kinematics law follows directly from the Maxwell equation (3.4) for $\vec{\rho}$:

$$\dot{\boldsymbol{\alpha}} = -\vec{\nabla} \times (\boldsymbol{\alpha} \times \vec{v}) \tag{3.5}$$

In the above, the following notations regarding cross products between a second-order tensor $\boldsymbol{T}$ and a vector $\vec{u}$ are used:

$$\boldsymbol{T} \times \vec{u} = \{T_{ij} u_k \in_{jkl}\} = \text{a } 3 \times 3 \text{ matrix with each row } i \text{ being the same row } i \text{ of } \boldsymbol{T} \text{ cross } \vec{u} \quad (3.6A)$$

$$\vec{u} \times \boldsymbol{T} = \{u_k T_{ij} \in_{jkl}\} = \text{a } 3 \times 3 \text{ matrix with each row } i \text{ being } \vec{u} \text{ cross the same row } i \text{ of } \boldsymbol{T} \quad (3.6B)$$

Exercise 3.1. Prove equation (3.5) using equation (3.4).

Solution: From equations (2.10A) and (3.4), we have

$$\dot{\alpha} = \vec{b} \otimes \dot{\vec{\rho}} = -\vec{b} \otimes [\vec{\nabla} \times (\vec{\rho} \times \vec{v})] = -\vec{\nabla} \times [\vec{b} \otimes (\vec{\rho} \times \vec{v})] = -\vec{\nabla} \times [(\vec{b} \otimes \vec{\rho}) \times \vec{v}] \quad (3.7)$$

The last step of equation (3.7) follows from equation (3.6A):

$$\left(\vec{b} \otimes \vec{\rho}\right) \times \vec{v} = \{b_i \rho_j v_k \epsilon_{jkl}\} = \{b_i (\vec{\rho} \times \vec{v})_l\} = \vec{b} \otimes (\vec{\rho} \times \vec{v})$$

where ϵ_{jkl} is the permutation symbol (= 1 if j, k and l are different and in clockwise rotation (123, 231, 312), = −1 if they are different and not in clockwise rotation (132, 321, 213), and = 0 if two or more of j, k and l are the same (112, 222, etc). Also, repeated occurrence of an index means summation.

The third step of equation (3.7) can be seen by letting $\vec{u} = \vec{\rho} \times \vec{v}$, and then

$$\vec{b} \otimes [\vec{\nabla} \times \vec{u}] = \{b_i (\vec{\nabla} \times \vec{u})_l\} = \{b_i \nabla_j u_k \epsilon_{jkl}\} = \{\nabla_j (b_i u_k) \epsilon_{jkl}\} = \left\{\nabla_j (\vec{b} \otimes \vec{u})_{ik} \epsilon_{jkl}\right\} = \vec{\nabla} \times \vec{b} \otimes \vec{u}$$

where the last step follows from equation (3.6B).

Equation (3.7) is therefore proven, which is just equation (3.5) since $\alpha = \vec{b} \otimes \vec{\rho}$.

Equation (3.5) can also be proven directly from the theory of deformation gradients to be discussed in section 5.6.1 in chapter 5. Referring to figure 5.7, the plastic deformation gradient is defined as a mapping $\mathbf{F^p} = \mathrm{d}\overline{\mathbf{X}}/\mathrm{d}\mathbf{X}$ that maps an infinitesimal line segment $\mathrm{d}\mathbf{X}$ in the undeformed state of the solid to the corresponding segment $\mathrm{d}\overline{\mathbf{X}}$ in the plastically deformed state. As shown in exercise 5.6 in chapter 5, $\alpha = \vec{\nabla}_{\mathbf{X}} \times \mathbf{F^p}$, which is one of Kröner's equations (Xia and El-Azab 2015). Furthermore, from $\mathbf{F^p}$, a plastic strain rate is given as $\dot{\mathbf{F}}^{\mathbf{p}} = \frac{\partial \dot{\overline{\mathbf{X}}}}{\partial \mathbf{X}}$, which may be equated to the strain rate $b\rho v\ (\hat{s} \otimes \hat{n})$ in the Taylor-Orowan equation (cf equation (1.1) and exercise 5.3), i.e., $\dot{\mathbf{F}}^{\mathbf{p}} = b\rho v\ (\hat{s} \otimes \hat{n})$. Since $\alpha = \vec{b} \otimes \vec{\rho}$ (equation (2.10a)) and $\vec{\rho} \times \vec{v} = -\rho v \hat{n}$ (figure 3.1), $\alpha \times \vec{v} = (\vec{b} \otimes \vec{\rho}) \times \vec{v} = \vec{b} \otimes (\vec{\rho} \times \vec{v}) = -b\rho v\ (\hat{s} \otimes \hat{n}) = -\dot{\mathbf{F}}^{\mathbf{p}}$. Hence, $\dot{\alpha} = \vec{\nabla}_{\mathbf{X}} \times \dot{\mathbf{F}}^{\mathbf{p}} = -\vec{\nabla}_{\mathbf{X}} \times (\alpha \times \vec{v})$ which is equation (3.5) (Xia and El-Azab 2015). Note that here, $\vec{\nabla}_{\mathbf{X}}$ is the del operator with respect to the coordinates $\mathbf{X}$ in the undeformed solid (figure 5.7), which is really $\vec{\nabla}$ in the present context.

Exercise 3.2 Show that with the kinematics law in equation (3.4), $\vec{\rho}$ always satisfies the divergence-free condition $\vec{\nabla} \cdot \vec{\rho} = 0$. What is the significance of this?

Solution: Let $\vec{u} = \vec{\rho} \times \vec{v}$. Then, from equation (3.4),

$$\vec{\nabla} \cdot \dot{\vec{\rho}} = -\vec{\nabla} \cdot (\vec{\nabla} \times \vec{u}) = -(\nabla_k \hat{e}_k) \cdot (\nabla_i u_j \epsilon_{ijk} \hat{e}_k) = -\nabla_k \nabla_i u_j \epsilon_{ijk}$$

Writing out the above in full,

$$\vec{\nabla} \cdot \dot{\vec{\rho}} = -(\nabla_3 \nabla_1 u_2 + \nabla_1 \nabla_2 u_3 + \nabla_2 \nabla_3 u_1 - \nabla_2 \nabla_1 u_3 - \nabla_1 \nabla_3 u_2 - \nabla_3 \nabla_2 u_1) = 0 \quad (3.8)$$

Therefore, if the initial $\vec{\rho}$ is divergence free, then integrating equation (3.8) will give a $\vec{\rho}$ that is also divergence free for all time t, i.e., $\vec{\nabla} \cdot \vec{\rho} = 0$. Divergence-free vector fields have no source, sink or disconnectivity, i.e., $\vec{\rho}$ represents connected dislocation lines with no termination, as must be the case for dislocations.

3.2 Other forms of Maxwell's kinematics law

3.2.1 Movement-tilting form

For a dislocation bundle moving at velocity $\vec{v}$ on a slip plane of normal $\hat{n}$ as shown in figure 3.2, its kinematics equation can be re-cast into a form that is much easier to understand.

Let $\hat{\xi}$ be the dislocation direction as shown in figure 3.2. Then, $\vec{v}$, $\hat{\xi}$ and $\hat{n}$ must be mutually perpendicular so we can write:

$$\vec{v} = v(\hat{n} \times \hat{\xi}), \quad \vec{\rho} = \rho\hat{\xi}, \text{ and } \vec{\rho} \times \vec{v} = \rho v \, \hat{\xi} \times (\hat{n} \times \hat{\xi}) = \rho v \hat{n}$$

Therefore, the kinematics law from equation (3.4) is:

$$\dot{\vec{\rho}} = -\vec{\nabla} \times (\vec{\rho} \times \vec{v}) = -\vec{\nabla} \times (\rho v \hat{n})$$

Now, choose the coordinate system to be the Cartesian along $\hat{\xi}$, $\hat{v} = \hat{\xi} \times \hat{n}$, and $\hat{n}$. Then, $\dot{\vec{\rho}} = -\vec{\nabla} \times (\rho v \hat{n}) = -(\hat{\xi}\, \nabla_{\hat{\xi}} + \hat{v}\nabla_{\hat{v}}) \times (\rho v \hat{n}) = -\nabla_{\hat{v}}\, (\rho v)\hat{\xi} + \nabla_{\hat{\xi}}\, (\rho v)\hat{v}$.

Note that $\nabla_{\hat{v}}\, (\rho v) = \vec{\nabla}\, (\rho v) \cdot \hat{v}$, and since $\vec{\rho}$ is divergence free from exercise 3.2, ρ is constant along $\hat{\xi}$, so that $\nabla_{\hat{\xi}}\, (\rho v) = \rho\, \nabla_{\hat{\xi}}\, v$. Therefore,

$$\dot{\vec{\rho}} = \underbrace{-[\vec{\nabla}\, (\rho v) \cdot \hat{v}]\hat{\xi}}_{\text{Movement}} + \underbrace{\rho(\nabla_{\hat{\xi}}\, v)\hat{v}}_{\text{tilt}} \quad (3.9)$$

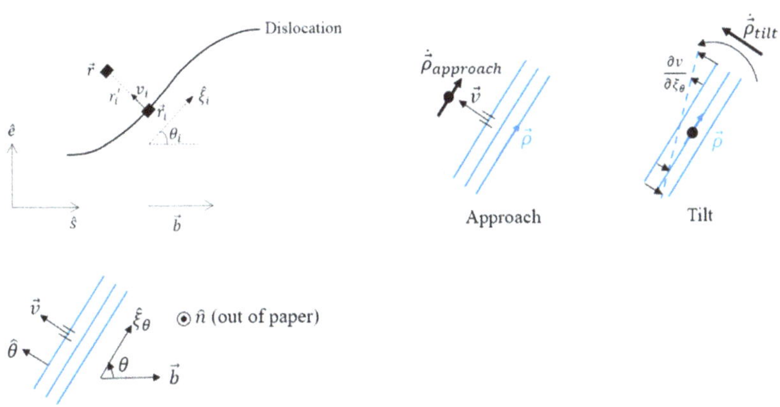

Figure 3.2. Dislocation moving on a slip plane, and definitions of $\hat{\theta}$, $\hat{\xi}_\theta$, $\hat{n}$, and $\vec{v}$.

The first term represents the rate of change of the dislocation density as dislocations approach a field point (i.e., an observer) in space, and the change recorded at the field point is along the dislocation line direction $\hat{\xi}$ because the approaching dislocations will augment density along this direction. In the second term in equation (3.9), $\nabla_{\hat{\xi}}\ v$ is simply the rotational speed of the dislocation bundle, and so this term represents the change in $\vec{\rho}$ as the dislocation is rotating; the vector change is along the $\hat{v}$ direction for positive value of $\nabla_{\hat{\xi}}\ v$, as shown in figure 3.2. Thus, the rotation only changes the direction but not the magnitude of the $\vec{\rho}$ vector, and a change in $\rho = |\vec{\rho}|$ is only due to the movement term in equation (3.9), i.e.,

$$\dot{\rho} = -\vec{\nabla}\,(\rho v)\cdot\hat{v} = -\frac{\partial(\rho v)}{\partial r_{\hat{\theta}}} \tag{3.9A}$$

The rotational speed $\nabla_{\hat{\xi}}\ v$ of the dislocation bundle is also the rate of change of the latter's character θ, i.e.,

$$\dot{\theta} = \nabla_{\hat{\xi}}\ v \tag{3.9B}$$

Equations (3.9A) and (3.9B) are therefore governing equations for ρ and θ, respectively, which if solved would fully specify the vector-density $\vec{\rho}$.

3.2.2 Divergence-curvature form

This form of the kinematics law for line density is of historical interest although it is of little practical use because of the coupling with the line curvature. Consider a bundle of curved dislocations with local curvature k near field point $\vec{r}$ (figure 3.3). Apply a cylindrical coordinate system $\vec{r} = (r, \phi)$ with the origin at the centre of curvature. For any vector field $\vec{J}(r, \phi) = J_r(r, \phi)\,\hat{r} + J_\phi(r, \phi)\,\hat{\phi}$ where $\hat{r}$ and $\hat{\phi}$ are unit vectors along the radial and circumferential directions, respectively, its divergence is given by:

$$\vec{\nabla}\cdot\vec{J} = \frac{1}{r}\frac{\partial(rJ_r)}{\partial r} + \frac{1}{r}\frac{\partial J_\phi}{\partial\phi}$$

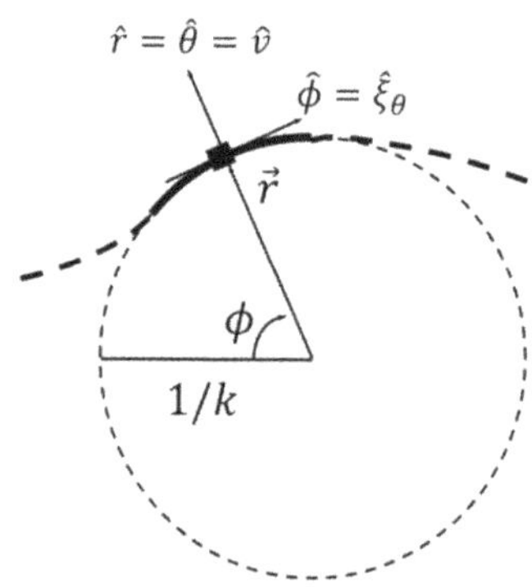

Figure 3.3. Curved dislocation bundle in cylindrical coordinates $\vec{r} = (r, \phi)$.

Now, set $\vec{J} = \rho\vec{v} = \rho v\hat{v} = \rho v\hat{r}$ to be the flux of dislocations, so that $J_r = \rho v$, $J_\phi = 0$. Therefore, we have

$$\vec{\nabla} \cdot \vec{J} = \vec{\nabla} \cdot (\rho\vec{v}) = \frac{1}{r}\frac{\partial(r\rho v)}{\partial r} = \frac{\rho v}{r} + \frac{\partial(\rho v)}{\partial r} = \rho v k + \frac{\partial(\varrho v)}{\partial r_{\hat{\theta}}} \tag{3.10}$$

In the above, $\rho v/r = \rho v k$ since ρ is non-zero only when $\vec{r} = (r, \phi)$ falls on the curved dislocation so that $r = 1/k$. Also, $r = r_{\hat{\theta}}$ because $\hat{r} = \hat{\theta} = \hat{v}$ from figure 3.3. Substituting the $\partial(\rho v)/\partial r_{\hat{\theta}}$ from equation (3.10) into equation (3.9A), we obtain

$$\dot{\rho} = -\vec{\nabla} \cdot (\rho\vec{v}) + \rho v k \tag{3.11}$$

which is another version of the kinematics law for lines, first derived by Sedláček *et al* (2003).

Despite equation (3.11) being equivalent to equation (3.9A), it involves the line curvature $k(\vec{r})$ for which another governing law needs to be available (Hochrainer *et al* 2007). The equivalent kinematics law in equation (3.9A) is neat as no curvature field is involved, and the only condition needed to solve it would be a dynamics law for the velocity field $\vec{v}$, but this would also be necessary if one wishes to solve equation (3.11). There is therefore hardly any advantage in working with equation (3.11).

Exercise 3.3 The curvature term $\rho v k$ in equation (3.11) is there to represent the notion that as a curved dislocation moves, new line length will be generated, which contributes to $\dot{\rho}$. Discuss how this aspect is covered in equation (3.9A), which does not have the $\rho v k$ term.

Solution: Consider a simple example in figure 3.4 where concentric dislocation loops are continuously sent out from the origin at a uniform speed. In this case, the

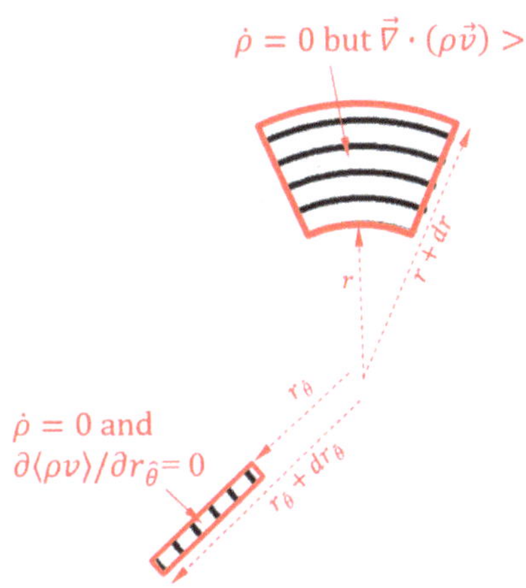

Figure 3.4. Concentric loops expanding from centre; $\dot{\rho} = 0$ except at centre.

only dislocation source is at the origin at which $\dot{\rho} > 0$; elsewhere, dislocations only move and so $\dot{\rho} = 0$.

In addition to the curvature term $\rho v k$, equation (3.11) also involves the divergence term $\vec{\nabla} \cdot (\rho \vec{v})$, which is a 2D differential operation on the flux $\vec{J} = \rho \vec{v}$ on the slip plane. Let us use the fan-shaped control volume in figure 3.4 to consider $\vec{\nabla} \cdot (\rho \vec{v})$: because the inner arc length where dislocations enter the control volume is shorter than the outer arc length where dislocations exit, and no dislocation enters the control volume on the two radial sides, this fan-shaped control volume will record a positive divergence, i.e., $\vec{\nabla} \cdot (\rho \vec{v}) > 0$. However, we have seen above that $\dot{\rho} = 0$ inside the control volume, and this is why the $\rho v k$ term, which is > 0, is needed to offset the $-\vec{\nabla} \cdot (\rho \vec{v})$ term in equation (3.11). So there is nothing wrong with equation (3.11), only that it is complicated to use.

How does equation (3.9A) work for the expanding loops in figure 3.4? Unlike the divergence $\vec{\nabla} \cdot (\rho \vec{v})$, $\partial(\rho v)/\partial r_{\hat{\theta}}$ is a linear operation on the flux ρv along the dislocation-normal direction $\hat{\theta}$. So the fan-shaped control volume in figure 3.4 is inapplicable, and we should use the slit control volume along the radial direction $\hat{\theta}$. With this slit control volume, we will be counting the same density flux arriving at the $r_{\hat{\theta}}$ end as that leaving at the $r_{\hat{\theta}} + dr_{\hat{\theta}}$ end, and so $\partial(\rho v)/\partial r_{\hat{\theta}} = 0$, matching the expected $\dot{\rho} = 0$. So equation (3.9A) applies equally well to the movement of curved dislocations.

In fact, the notion of 'movement of curved dislocations produces new line length' needs to be interpreted with extreme care. While new line length is generated, the dislocation density (line length threading unit cross-sectional area) should remain the same as curved dislocations move. Therefore, $\dot{\rho} = 0$ should not be confused with no line length generated—as curved dislocations move, they cover more of the slip plane so both the total line length and the volume in which dislocations reside increase, thus keeping the density constant. This explains why in the case of the expanding loops in figure 3.4, $\rho v k > 0$, $\vec{\nabla} \cdot (\rho \vec{v}) > 0$, and yet $\dot{\rho} = 0$.

Exercise 3.4 In earlier discussions of CDD in the literature, movement of dislocations was considered as analogous to diffusion of particles (e.g., the Walgraef–Aifantis reaction-diffusion models (Walgraef and Aifantis 1985)). Is this analogy valid?

Solution: Diffusion of particles is governed by Fick's second law in equation (2.2):

$$\frac{\partial C}{\partial t} = -\vec{\nabla} \cdot \vec{J} \tag{2.2}$$

The said analogy would imply that for dislocations, (i) $\vec{J} = \rho \vec{v}$, and (ii) $\dot{\rho} = -\vec{\nabla} \cdot \vec{J} = -\vec{\nabla} \cdot (\rho \vec{v})$. However, we have seen from exercise 3.3 that the last equation is not correct: for the expanding loops in figure 3.4, $\vec{\nabla} \cdot (\rho \vec{v}) > 0$ but $\dot{\rho} = 0$. Rather, $\dot{\rho} = -\partial(\rho v)/\partial r_{\hat{\theta}} = -\vec{\nabla} \cdot (\rho \vec{v}) + \rho v k$, even if rotations are ignored.

Therefore, the kinematics laws for particles and lines (dislocations) are distinctively different with little analogy between them. Dislocations are connected lines, so their correct kinematics law has to have the line connectivity condition (i.e. $\vec{\nabla} \cdot \vec{\rho} = 0$, see exercise 3.2) satisfied. Also, line rotation needs to be considered in the kinematics law as well (see equation (3.9B)). No such connectivity or rotation condition is required for the movement of particles. We are used to calling transport phenomena satisfying equation (2.2) diffusion, so it is better not to call dislocation movement diffusion, in order to avoid confusion. Therefore, dislocations do not diffuse!

3.2.3 Finite-difference form

Here, we introduce another form of the kinematics law that is particularly useful for developing a numerical solver. Consider a set of dislocations of character θ moving (without tilting) across a density-measurer surface $\Delta\vec{s}$ in figure 3.5. During time δt:

$$\text{lines passing into } \Delta\vec{s} \text{ via point A (situated at } \vec{r}_A) = (\vec{\rho} \times \vec{v})|_{\vec{r}_A}\, \delta t \cdot b\hat{n}$$

$$\text{lines leaving } \Delta\vec{s} \text{ via point B (situated at } \vec{r}_A + \Delta\vec{s}) = (\vec{\rho} \times \vec{v})|_{\vec{r}_A + \Delta\vec{s}}\, \delta t \cdot b\hat{n}$$

where $\hat{n}$ and b are the unit normal and thickness of the slip plane, respectively, and $\Delta\vec{s}$ is the vector of the intersecting line between the measurer surface and the slip

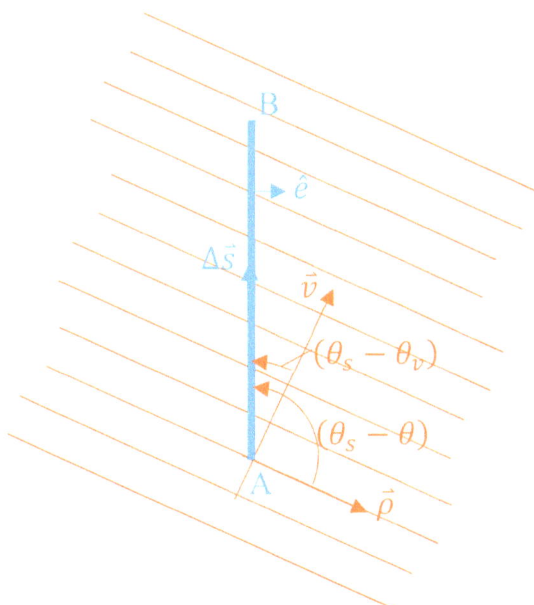

Figure 3.5. Dislocations (in orange colour) of line density $\vec{\rho}$ and dislocation character θ moving with velocity $\vec{v}$ on a slip plane (the plane of this paper) with atomic thickness b. A surface normal to and intersecting the slip plane along $\Delta\vec{s} = \overrightarrow{AB}$ is used to count the dislocations threading it. θ_s, θ_v and θ are angles measured anticlockwise from a given reference direction (not shown for clarity) to the line $\Delta\vec{s}$, $\vec{v}$ and $\vec{\rho}$, respectively.

plane. Let $\hat{e}$ be the normal direction of the measurer surface, the dislocation density as per the definition in equation (2.9) is $\vec{\rho} \cdot \hat{e} = \rho\hat{\xi} \cdot \hat{e}$ where ρ is the magnitude of $\vec{\rho}$ and $\hat{\xi}$ its direction. Therefore, the net increase in the number of dislocations threading $\Delta\vec{s}$ during δt is:

$$\delta\rho\ \hat{\xi} \cdot \hat{e}\ b\Delta s = -[(\vec{\rho} \times \vec{v})|_{\vec{r}_A + \Delta\vec{s}} - (\vec{\rho} \times \vec{v})|_{\vec{r}_A}]\ \delta t \cdot b\hat{n}$$

where $\Delta s = |\Delta\vec{s}\,|$. It is easily seen that $\hat{\xi} \cdot \hat{e}\ \Delta s = (\hat{\xi} \times \Delta\vec{s}) \cdot \hat{n}$; hence the above equation becomes

$$\dot{\rho}\ \hat{\xi} \times \Delta\vec{s} = -\Delta(\vec{\rho} \times \vec{v})|_{\Delta\vec{s}} \tag{3.12}$$

where the change $\Delta(\vec{\rho} \times \vec{v})|_{\Delta\vec{s}}$ means that it is over a change $\Delta\vec{s}$ in the $\vec{r}$-space from point A. Considering only the vector magnitudes on both sides of equation (3.12), we have

$$\dot{\rho}\Delta s \sin(\theta_s - \theta) = -\Delta(\rho v)|_{\Delta\vec{s}} \tag{3.13}$$

where $(\theta_s - \theta)$ is the angle between $\Delta\vec{s}$ and $\hat{\xi}$ (see figure 3.5). It can be easily seen that equation (3.13) is in fact identical to equation (3.9A), since from figure 3.5, the distance $\Delta r_{\hat{\theta}}$ along the velocity direction is simply $\Delta r_{\hat{\theta}} = \Delta s \sin(\theta_s - \theta)$. Although equations (3.12) and (3.13) are derived by considering only the translation but not rotation of the dislocations, as noted in equation (3.9), rotation changes only the direction but not the magnitude ρ of $\vec{\rho}$, and so equations (3.12) and (3.13) are valid even though the dislocations also rotate within $\Delta\vec{s}$. Equation (3.13), together with equation (3.9B), is the finite-difference form of the kinematics law for $\vec{\rho}$ that we intend to seek.

3.3 Finite-difference numerical scheme

With equation (3.13), we are now ready to develop a numerical scheme for solving the dislocation density evolution. Figure 3.6 shows the slip plane discretized into a square mesh. Each mesh point (*i,j*) is the centre of two orthogonal counting surfaces $\overrightarrow{AB}$ and $\overrightarrow{CD}$ of the type shown in figure 3.5. Then, application of equation (3.13) to the counting surfaces $\overrightarrow{AB}$ and $\overrightarrow{CD}$ gives:

$$\dot{\rho}(\theta)|_{ABCD}\ \Delta s \cos(\theta_{\overrightarrow{AB}} - \theta_v) = -\Delta(\varrho v)|_{\overrightarrow{AB}} \tag{3.14}$$

$$\dot{\rho}(\theta)|_{ABCD}\ \Delta s \cos(\theta_{\overrightarrow{CD}} - \theta_v) = -\Delta(\varrho v)|_{\overrightarrow{CD}} \tag{3.15}$$

where $(\theta_{\overrightarrow{AB}} - \theta_v)$ and $(\theta_{\overrightarrow{CD}} - \theta_v)$ are the angles from the velocity direction $\vec{v}$ to $\overrightarrow{AB}$ and $\overrightarrow{CD}$, respectively (see figures 3.5 and 3.6), and $\Delta s = \lceil\overrightarrow{AB}\rceil = |\overrightarrow{CD}|$. Equations (3.14) and (3.15) can be merged to give:

$$\dot{\rho}|_{ABCD}\ \Delta s = -\Delta(\rho v)|_{\overrightarrow{AB}} \cos(\theta_{\overrightarrow{AB}} - \theta_v) - \Delta(\rho v)|_{\overrightarrow{CD}} \cos(\theta_{\overrightarrow{CD}} - \theta_v) \tag{3.16}$$

where $\Delta(\rho v)|_{\overrightarrow{AB}} = (\rho v)|_B - (\rho v)|_A$ and $\Delta(\rho v)|_{\overrightarrow{CD}} = (\rho v)|_D - (\rho v)|_C$.

Furthermore, the rotation of the dislocation is solved by integrating equation (3.9B) as follows:

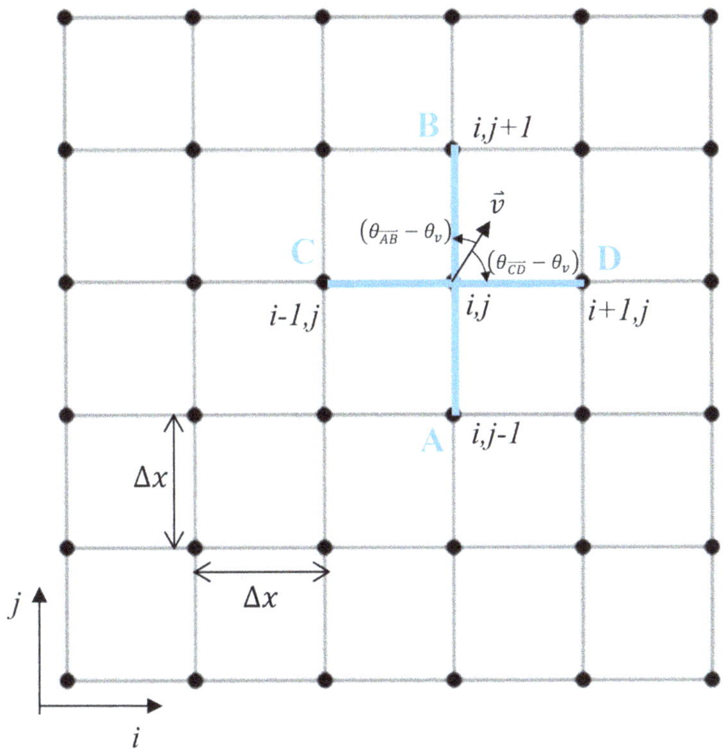

Figure 3.6. Discretization of slip plane (the plane of this page) into a square mesh. Each mesh point, say, (*i,j*) is the centre of two orthogonal counting surfaces AB and CD of the type shown in figure 3.5.

$$\theta = \int_0^t \frac{\partial v}{\partial \xi} \mathrm{d}t \tag{3.17}$$

where ξ is the spatial coordinate along the dislocation line direction. Note that for the square-mesh discretization scheme in figure 3.6, equation (3.16) may be explicitly written as:

$$\dot{\rho}\,|_{i,j}\,\Delta x = \frac{\left[(\rho v)\,|_{i,j-1} - (\rho v)\,|_{i,j+1}\right]}{2}\cos\left(\theta_{\overline{AB}} - \theta_v\right) + \frac{\left[(\rho v)\,|_{i-1,j} - (\rho v)\,|_{i+1,j}\right]}{2}\cos\left(\theta_{\overline{CD}} - \theta_v\right) \tag{3.16a}$$

where $\Delta x = \Delta s/2$ is the mesh size. Equations (3.16) or (3.16A), and equation (3.17), form the coupled numerical scheme for solving the kinematics of an intensive microstructure specified by the dislocation density field $\vec{\rho}(\vec{r})$.

3.4 Simple MATLAB codes for line-density kinematics

In this section, we present simple MATLAB codes for simulating moving lines.

3.4.1 Setting up line and coarse-graining into density

The first task is to coarse-grain the simulated line into a density function $\vec{\rho}(\vec{r})$, or the coupled functions $\rho(\vec{r})$ and $\theta(\vec{r})$, defined over the $\vec{r}$-space of the slip plane. To do this, the following steps can be used:

- *Setting up the slip-plane mesh*—the mesh points are stored in a matrix 'meshpoints', i.e., for the kth mesh point,

```
meshpoints(k,1)= x-coordinate of mesh point k;
meshpoints(k,2)= y-coordinate of mesh point k;
```

- *Setting up the initial line*—the points on the initial line to be simulated (which do not need to overlap with any of the meshpoints) are stored in the matrix 'line':

```
line = [lx;ly];
```

  Here, lx and ly are vectors of the x- and y-coordinates of the points on the line.
- *Finding the shortest distance of each mesh point to the line*—the external subroutine 'distance2curve' (D'Errico 2010) is used for this purpose:

```
[xy, distance, t_a] = distance2curve(line.',
meshpoints,'linear');
```

  Here, matrix 'distance', which lists the shortest distance to the curve from each of the input meshpoints is our desired output. Output 't_a', which lists the fractional arc lengths along the interpolating curve to the points closest to the input meshpoints, is used to ascribe the dislocation character to the meshpoints. Output 'xy' lists the closest points on the curve to the meshpoints, but this is not used for our purpose. On the input side, 'linear' is the selected method for the subroutine, and 'line.' is just the transpose of 'line' as the required input format.
- *Setting up the density distribution $\rho(\vec{r})$*—the 'distance' matrix from the step above gives the shortest distance $r_{i,j}$ of each meshpoint (i,j) to the line (see figure 3.7). Then, the density at point (i,j) is set up according to a Gaussian distribution with standard deviation σ:

$$\rho_{i,j} = Q(i,j) = \frac{1}{\sigma\sqrt{2\pi}} \exp\left(-\frac{r_{i,j}^2}{2\sigma^2}\right). \tag{3.18}$$

  The densities $\rho_{i,j}$ are stored in the matrix $Q(i,j)$.
- *Setting up the dislocation character distribution $\theta(\vec{r})$*—For each meshpoint (i,j) on the slip plane, its dislocation character θ is assigned to be the slope at the point on the curve closest to (i,j), as shown in figure 3.7. To do this, the 'line' defined above is first curve-fitted by the MATLAB function 'cscvn' and then

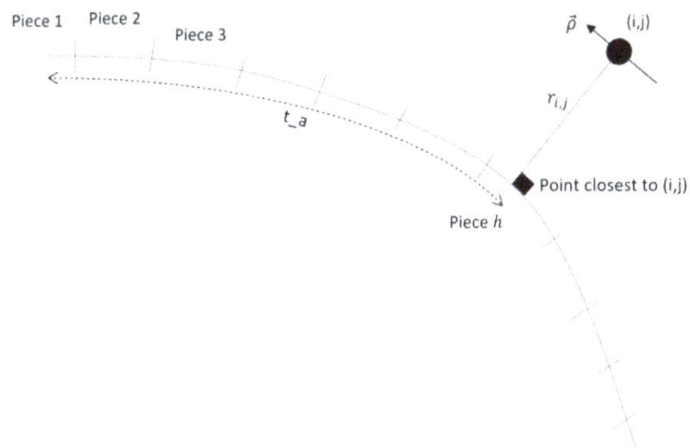

Figure 3.7. Ascribing *Q*(*i,j*) and theta(*i,j*) for the density $\rho(\vec{r})$ and character $\theta(\vec{r})$ distribution, respectively.

differentiated by function 'fnder'. The arc length 't_a' returned from the 'distance2curve' subroutine above is used to identify the point closest to (i,j) as well as the piece of the fitted curve (labelled by *h* in figure 3.7 and 'index' in the codes) containing this point, so that the fitted derivative coefficients from the 'fnder' function can be identified and used. The dislocation character function $\theta(\vec{r})$ is stored in the matrix theta(i,j) in the programme.

After setting up the $\rho(\vec{r})$ and $\theta(\vec{r})$ functions over space, the scheme explained in section 3.3 above can be used to simulate line kinematics.

3.4.2 A basic MATLAB programme for expanding circle

To illustrate the basic points discussed in sections 3.3 and 3.4.1, a circular dislocation loop expanding at a constant velocity will be a good example. The following codes are presented for this purpose.

3.4.2.1 Main programme

```
%% DDFD_discretek: for expanding circle kinematics
%% Preset parameters
b = 1;              % working length scale
dt = 1;             % time step
T = 50*dt;          % total time, 40 here is the number of time steps
Vsize = 10*b/dt;    % dislocation velocity
nxintervals = 101;  % number of mesh intervals in slip plane
magnitude=2000*b;   % size of simulation cell (slip plane)
thetas = 0;         % start angle of dislocation arc
thetaf = 2*pi;      % finish angle of dislocation arc
sdev= 50*b;         % standard deviation in gaussian profile of dislocation

%% Setting up square mesh
x=linspace(0,magnitude,nxintervals); % x is a vector of evenly spaced numbers
y=x;
```

```
dx=(x(end)-x(1))/(size(x,2)-1); % dx = mesh size; size(x,2) is the size of vector x
meshpoints=zeros((size(x,2))^2,2);
k=1;
for i=1:size(x,2)
    for j=1:size(y,2)
        meshpoints(k,1) = x(1,i);
        meshpoints(k,2) = y(1,j);
        k=k+1;
    end
end

%% Setting up the dislocation loop
xCenter = magnitude/2; % circle centre in the meshpoint coordinate system
yCenter = magnitude/2;
radius = .1*magnitude; % radius of the initial loop
nthetadis = floor((thetaf-thetas)/0.01)+1;
lx = zeros(1,nthetadis);
ly = zeros(1,nthetadis);
i = 1;
for thetadis = thetas : 0.01 : thetaf
    lx(1,i) = radius*cos(thetadis) + xCenter;
    ly(1,i) = radius*sin(thetadis) + yCenter;
    i = i+1;
end
line=[lx;ly];  % a matrix for the discrete points of the circle

%% Coarse-graining line into density function with a Gaussian cross-sectional profile
Q=zeros(size(x,2),size(y,2)); % Q(i,j) is the density over spatial domain (i,j).
QQ=Q; % QQ = 2D matrix for working density (rho), and is here initialized to Q


theta = zeros(size(x,2),size(y,2)); % theta = dislocation character (from -XX axis)
[xy,distance,t_a]=distance2curve(line.',meshpoints,'linear'); % see section 3.4.1
[QQ,theta]= discoarse(distance,t_a,sdev,x,line,Q,theta);

%% Preparing for integration
for i=1:size(x,2)
    for j=1:size(y,2)
        Vbydx(i,j)=vsize/dx;  % set constant velocity v at all points, vbydx=v/dx
    end
end

%% Integration
for time=1:round(T/dt)
    QQ = difff1(QQ,dt,Vbydx,theta);   % Runge-Kutta integration subroutine
    %
    % Density QQ can be plotted here, which is skipped for clarity purpose
    %
end
```

3.4.2.2 Subroutine for coarse-graining line into density with Gaussian cross-sectional profile

```
function [Q,theta] = discoarse(distance, t_a, sdev, x, line, Q, theta)
% This function obtains the density according to the distance of nodes to a
dislocation line in the form of Z, a 2D matrix.
    h_a=reshape(t_a.',[size(x,2),size(x,2)]);
% reshape t_a (a vector size(x)^2 long for the meshpoints) into a size(x) by size (x)
matrix
    distance=reshape(distance.',[size(x,2),size(x,2)]);
% reshape distance (a vector size(x)^2 long for the meshpoints) into a size(x) by
size(x) matrix
    Z = 1/(sdev*sqrt(2*pi))*exp(-1*((distance).^2)/(2*sdev^2));
% since distance is a size(x) by size(x) matrix, so is Z, which is now the Gaussian
profile of the dislocation line over the meshpoint space.
    Z(isnan(Z))=0; % setting the NaN elements of Z to zero
    t = cumsum([0;((diff(line.').^2)*ones(2,1)).^(1/4)]).'; % t is cumulative sum of
squares of differences of successive points in "line"
    derivation=fnder(cscvn(line));  % derivative of curve
    [Q,theta]=iter(Z,t,derivation,Q,h_a,theta);
end

function [Q,theta]=iter(Z,t,derivation,Q,h_a,theta)
% This function converts Z into density Q and character theta which are 2D matrices
over space.
    for p=1:size(h_a,1)     % size(h_a,1) = size(x), size(h_a,2) = size(x)
        for l=1:size(h_a,2) % (p,l) are indices over the meshpoints
        if ~isnan(h_a(p,l)) % if h_a(p,l) is not NaN
            if h_a(p,l)>1   % h_a is fractional arc length to the closest point
               h_a(p,l)=1;  % capping any h_a > 1 to 1
            end
            index = find((t/t(end))>=(h_a(p,l)), 1, 'first');
      % index = the fitted curve piece containing point on curve closest to meshpoint
            if index==1
               tangent=[-dot(derivation.coefs((2*index)-1,:),[0,0,1]),


                                    dot(derivation.coefs((2*index),:),[0,0,1])];
            elseif ~isempty(index)
               t0=(h_a(p,l)*t(end))- t(index-1);
               tangent=[-dot(derivation.coefs((2*index)-3,:),[t0^2,t0^1,1]),
                                    dot(derivation.coefs((2*index)-2,:),[t0^2,t0^1,1])];
            end
            theta(l,p)=2*pi-tan2thet(tangent);
% theta is dislocation character angle (0 to 2*pi) measured anticlockwise from XX.
% It is also negative of the angle from the velocity direction to YY axis.
            Q(l,p)=Q(l,p)+ Z(p,l);  % For Q and theta, i is horizontal, j vertical
        end
        end
    end
end
```

3.4.2.3 Runge–Kutta integration subroutine

```
function [FQ]= difff1(QQ,dt,Vbydx,theta)
G=QQ;
K1=FDM1(G,Vbydx,theta);
G=QQ+K1*(dt/2);
G(G<0)=0;
K2=FDM1(G,Vbydx,theta);
G=QQ+K2*(dt/2);
G(G<0)=0;
K3=FDM1(G,Vbydx,theta);
G=QQ+K3*dt;
G(G<0)=0;
K4=FDM1(G,Vbydx,theta);
FQ=QQ+((K1+(2*K2)+(2*K3)+K4)*dt)/6;
FQ(FQ<0)=0;
end

function [K]=FDM1(G,Vbydx,theta)
K=zeros(size(G,1),size(G,2));
for i=2:size(G,2)-1
    for j=2:size(G,1)-1
        GA=G(i,j-1)/2; % A
        GB=G(i,j+1)/2; % B
        GC=G(i-1,j)/2; % C
        GD=G(i+1,j)/2; % D
        K(i,j)=-Vbydx(i,j)*((GB-GA)*cos(-theta(i,j))+(GD-GC)*sin(-theta(i,j)));
                                                                    % Eqn.(3.16)
    end
end
end
```

3.4.2.4 Simulated results

Figure 3.8 shows results of the expanding-loop simulation using the codes above. The total dislocation content $\sum_{(i,j)}\rho(i, j)\Delta x^2$ increases proportionally with the loop radius as expected.

3.4.3 Further considerations for solving line-density kinematics

3.4.3.1 The rhombus element

In figure 3.8(a), the loop simulated using the simple square mesh in figure 3.6 does not maintain its perfect sphericity and uniform density as it expands. This is due to the crudeness of the square mesh, i.e., in equation (3.16A), $\dot{\rho}|_{i,j}$ at each node *(i,j)* is evaluated from the flux at only four of its nearest neighbours *(i,j−1)*, *(i,j+1)*, *(i+1, j)* and *(i−1,j)*. Here, we introduce another mesh, namely, the rhombus mesh, that enables the $\dot{\rho}$ at each node to be evaluated from more of its neighbours, so as to improve accuracy.

The rhombus mesh is illustrated in figure 3.9. Here, each nodal point *(i,j)* etc. of the mesh is at the centre of a pixel-element such as the beige-coloured rhombus ABCD. The red grid lines in figure 3.9 therefore mark the boundaries of the pixels, and compared with figure 3.6, the *i*–*j* axes are now rotated 45° clockwise. In each pixel, say, the ABCD in figure 3.9, two orthogonal surfaces $\overrightarrow{AB}$ and $\overrightarrow{CD}$ are used to count

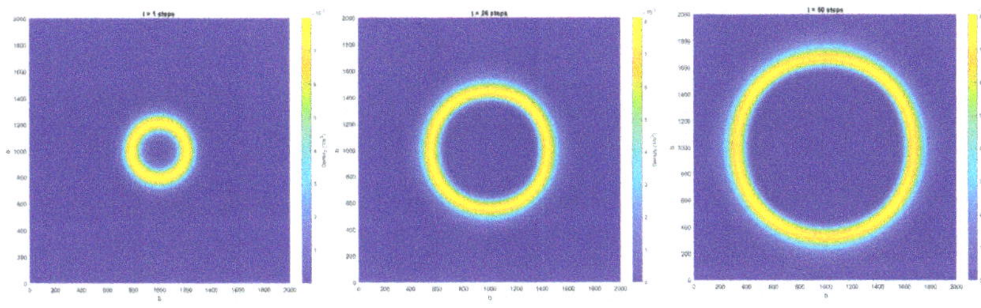

(a)

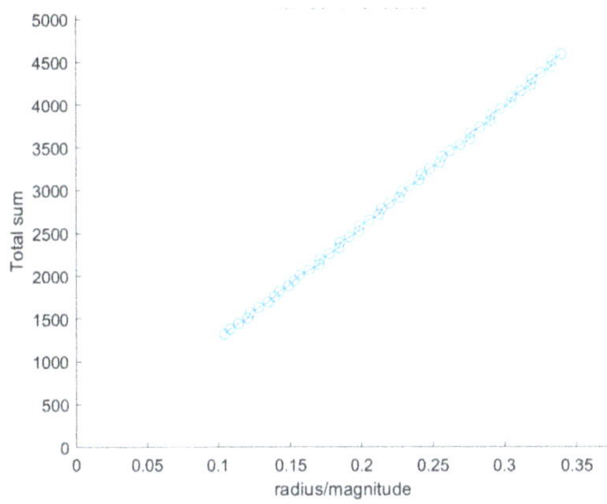

(b)

Figure 3.8. (a) Snapshots of expanding loop simulated using the codes in section 3.4.2 with the cubic mesh. (b) Growth of total dislocation content, $\sum_{(i,j)}\rho(i,j)\Delta x^2$, vs normalized radius of the loop during expansion.

dislocations as before, and hence equation (3.16) remains valid. Instead of equation (3.16A) for the cubic element where each end point A, B, etc. of the counting surfaces is a mesh point such as *(i,j-1)* or *(i,j+1)*, for the rhombus element in figure 3.9, the end points *A*, *B*, *C* and *D* are not mesh points but are surrounded by mesh points, at which density, character and velocity are registered. Therefore, the flux at *A*, *B*, *C* and *D* can be obtained by averaging from the surrounding mesh points as follows:

$$\begin{aligned}
(\rho v)|_A &= \frac{1}{4}[(\rho v)|_{i,j} + (\rho v)|_{i+1,j} + (\rho v)|_{i,j-1} + (\rho v)|_{i+1,j-1}] \\
(\rho v)|_B &= \frac{1}{4}[(\rho v)|_{i-1,j+1} + (\rho v)|_{i,j+1} + (\rho v)|_{i,j} + (\rho v)|_{i-1,j}] \\
(\rho v)|_C &= \frac{1}{4}[(\rho v)|_{i-1,j} + (\rho v)|_{i,j} + (\rho v)|_{i,j-1} + (\rho v)|_{i-1,j-1}] \\
(\rho v)|_D &= \frac{1}{4}[(\rho v)|_{i,j+1} + (\rho v)|_{i+1,j+1} + (\rho v)|_{i+1,j} + (\rho v)|_{i,j}]
\end{aligned} \tag{3.19}$$

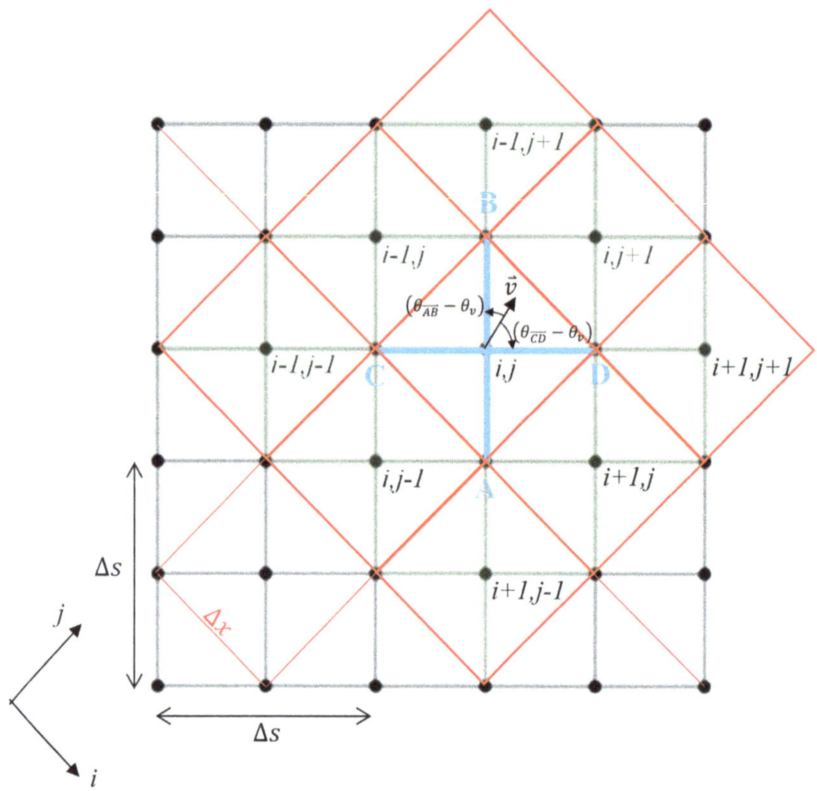

Figure 3.9. The rhombus mesh with improved nodal connectivity. Computed quantities (density, dislocation character, velocity) are stored at mesh points (*i,j*), etc, which are centres of rhombus pixel-elements such as ABCD, where AB and CD are two orthogonal counting surfaces for dislocation lines. Adapted from Kalaei *et al* (2022).

Substituting equation (3.19) into equation (3.16), we have

$$\begin{aligned}\dot{\rho}\big|_{i,j} = \frac{1}{4\,\Delta s}\Big\{&\Big[(\rho v)\big|_{i+1,j} + (\rho v)\big|_{i,j-1} + (\rho v)\big|_{i+1,j-1} - (\rho v)\big|_{i-1,j+1} - (\rho v)\big|_{i,j+1} \\ &- (\rho v)\big|_{i-1,j}\Big]\cos(\theta_{\overline{AB}} - \theta_v) \\ &+\Big[(\rho v)\big|_{i-1,j} + (\rho v)\big|_{i,j-1} + (\rho v)\big|_{i-1,j-1} - (\rho v)\big|_{i,j+1} - (\rho v)\big|_{i+1,j+1} \\ &- (\rho v)\big|_{i+1,j}\Big]\cos(\theta_{\overline{CD}} - \theta_v)\Big\}\end{aligned} \tag{3.20}$$

where $\Delta s = |\overline{AB}| = |\overline{CD}| = \sqrt{2}\,\Delta x$, Δx is the mesh size. Equation (3.20) is the numerical scheme for the rhombus mesh, and compared with equation (3.16A) for the cubic mesh, the $\dot{\rho}$ at (i,j) is evaluated from the fluxes at $(i \pm 1, j \pm 1)$, i.e., not only its four nearest neighbours but also the four second-nearest neighbours.

The rhombus element can be executed by the following function 'difff2' which replaces the 'difff1' in section 3.4.2.3. Since the second neighbours are involved for each mesh node, periodic boundary conditions are applied (via function '`periodic_2d`') to make sure that boundary nodes have proper neighbours connected:

```
function [FQ]= difff2(Q,dt,Vbydx,theta)
% This version of the difff function executes the rhombus element in Figure 3.9
G=Q;
cos_thetaV = cos(theta+pi/4);  % thetaV is dislocation normal from CD axis, theta is
dislocation line inclination angle from XX axis which is along i.
sin_thetaV = sin(theta+pi/4);  % CD axis is 45 deg. from XX axis.
K1=FDM(G,Vbydx,cos_thetaV,sin_thetaV, Q, dt/2);
G=Q+K1*dt/2;
G(G<0)=0;
K2=FDM(G,Vbydx,cos_thetaV,sin_thetaV, Q, dt/2);
G=Q+K2*dt/2;
G(G<0)=0;
K3=FDM(G,Vbydx,cos_thetaV,sin_thetaV, Q, dt);
G=Q+K3*dt;
G(G<0)=0;
K4=FDM(G,Vbydx,cos_thetaV,sin_thetaV, Q, dt);
FQ=Q+(K1+2*K2+2*K3+K4)*dt/6;
FQ(FQ<0)=0;
%FQ=Q+K1*dt;  % simple FDM
end

function [K]=FDM(G,Vbydx,cos_thetaV,sin_thetaV, Q, dt)
K=zeros(size(G,1),size(G,2));
px=size(G,1); py=size(G,2);
[G_b]=periodic_2d(G, size(G,1), size(G,2), 1, px, py);
[Vbydx_b]=periodic_2d(Vbydx, size(Vbydx,1), size(Vbydx,2), 1, px, py);
for i=1:size(G,1)
    for j=1:size(G,2) % c.f. eqn. (3.19)
        flux_A=(G_b(px+i,py+j-1)*Vbydx_b(px+i,py+j-1)+G_b(px+i+1,py+j-
1)*Vbydx_b(px+i+1,py+j-1)+G_b(px+i+1,py+j)*Vbydx_b(px+i+1,py+j))/4; % A
        flux_B=(G_b(px+i,py+j+1)*Vbydx_b(px+i,py+j+1)+G_b(px+i-
1,py+j+1)*Vbydx_b(px+i-1,py+j+1)+G_b(px+i-1,py+j)*Vbydx_b(px+i-1,py+j))/4; % B
        flux_C=(G_b(px+i-1,py+j-1)*Vbydx_b(px+i-1,py+j-1)+G_b(px+i,py+j-
1)*Vbydx_b(px+i,py+j-1)+G_b(px+i-1,py+j)*Vbydx_b(px+i-1,py+j))/4; % C
flux_D=(G_b(px+i+1,py+j+1)*Vbydx_b(px+i+1,py+j+1)+G_b(px+i,py+j+1)*Vbydx_b(px+i,py+j+
1)+G_b(px+i+1,py+j)*Vbydx_b(px+i+1,py+j))/4; % D
        K(i,j)=-((flux_B-flux_A)*sin_thetaV(i,j)+(flux_D-
flux_C)*cos_thetaV(i,j))/sqrt(2); % thetaV is dislocation normal from CD axis, c.f.
eqn. (3.16), sqrt2 for converting Vbydx to Vbyds c.f. Fig. 3.9
        if Q(i,j)+K(i,j)*dt<0
            K(i,j)=-Q(i,j)/dt;
        end
    end
end
end

function [mat_p]=periodic_2d(mat, m, n, q, px, py)
if n>=px && m>=py
    mat_p=zeros(m+px*2,n+py*2,q);

    mat_p(1+px:m+px,1+py:n+py,:)=mat;

    % Left side and bottom left corner
    mat_p(1:px,1+py:n+py,:)=mat(m-px+1:m,1:n,:);  % left side
    mat_p(1:px,1:py,:)=mat(m-px+1:m,n-py+1:n,:);  % bottom left corner
```

```
        % Top left corner and top side
        mat_p(1+px:m+px,n+py+1:n+2*py,:)=mat(1:m,1:py,:); % top side
        mat_p(1:px,n+py+1:n+2*py,:)=mat(m-px+1:m,1:py,:); % top left corner

        % Top right corner and right side
        mat_p(m+px+1:m+2*px,1+py:n+py,:)=mat(1:px,1:n,:); % right side
        mat_p(m+px+1:m+2*px,n+py+1:n+2*py,:)=mat(1:px,1:py,:); % top right corner

        % Bottom right corner and bottom side
        mat_p(1+px:m+px,1:py,:)=mat(1:m,n-py+1:n,:); % bottom side
        mat_p(m+px+1:m+2*px,1:py,:)=mat(1:px,n-py+1:n,:); % bottom right corner
    end
    end
```

Figure 3.10 shows the snapshots and total dislocation content simulated using the main programme in section 3.4.2.1 and the difff2 function above. With the rhombus element, the expanding loop exhibits improved sphericity and uniformity in density than that with the cubic element in figure 3.8.

3.4.3.2 Movement-tilting synchronization

The expanding circular loop considered above does not involve rotation, as each line segment only moves outward from the origin and does not tilt, i.e., in equation (3.9B), $\dot{\theta} = 0$. For simulations involving line rotation, the integration in equation (3.17)

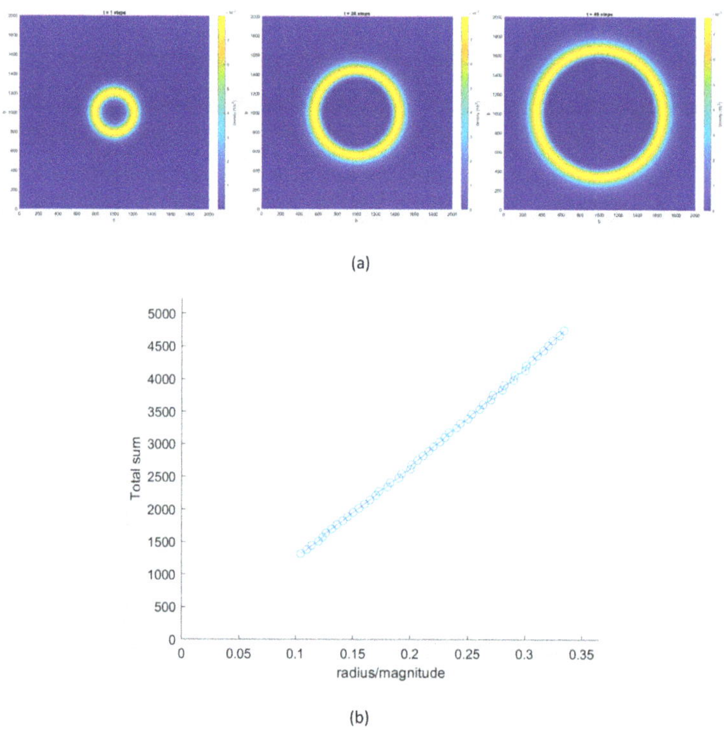

Figure 3.10. (a) Snapshots of expanding loop simulated using the difff2 function with the rhombus mesh. (b) Growth of total dislocation content, $\sum_{(i,j)} \rho(i,j)\Delta x^2$, vs normalized radius of the loop during expansion.

should be carried out, and to do this, the line orientation $\hat{\xi}$ needs to be established for each small line segment, for the rotational speed $\partial v/\partial \xi$ to be calculated along the line coordinate ξ. In the following 'rotation1' subroutine, $\hat{\xi}$ is established by considering the dislocation density within a neighbourhood of $\pm$ nb cells from the current node, and integration to find the rotation is carried out by the Runge–Kutta method:

```
function [theta,rot]=rotation1(theta,QQ,V,dt,dx,dy,dis_width,rho_min)
% This function finds the rotation rot during dt
xi = zeros(size(V,1),size(V,2),2);  % direction cosines of GND at each site
xi(:,:,1) = cos(theta(:,:));  % projection of GND onto screw axis
xi(:,:,2) = sin(theta(:,:));
d_width=ceil(dis_width/dx);
nb=5*d_width;   % neighbour distances within which velocity gradient is calculated

thet=theta;
K1=theta_diff(thet,QQ,V,rho_min,xi,dx,dy,d_width,nb);

% R-K integration
thet=theta+K1*dt/2;
K2=theta_diff(thet,QQ,V,rho_min,xi,dx,dy,d_width,nb);
thet=theta+K2*dt/2;
K3=theta_diff(thet,QQ,V,rho_min,xi,dx,dy,d_width,nb);
thet=theta+K3*dt;
K4=theta_diff(thet,QQ,V,rho_min,xi,dx,dy,d_width,nb);
rot=(K1+2*K2+2*K3+K4)/6*dt;
theta=theta+rot;
end

function [dVbydxi]=theta_diff(thet,QQ,V,rho_min,xi,dx,dy,d_width,nb)
dVbydxi=zeros(size(thet,1),size(thet,2));
for i=1:size(thet,1)
    for j=1:size(thet,2)
        if QQ(i,j)<=rho_min
            continue
        end
        dVbydxi(i,j)=0;
        n_count=0;
        for ii=i-nb:i+nb
            for jj=j-nb:j+nb
                if ii>0 && ii<=size(V,1) && jj>0 && jj<=size(V,2) && ii~=i && jj~=j
                    if QQ(ii,jj)<=rho_min
                        continue
                    end
                    d = abs(-(ii-i)*xi(i,j,2)+(jj-j)*xi(i,j,1));
                        % distance along current disln normal direction at (i,j)
                    if d>d_width
                        continue
                    end
                    d_xi = (ii-i)*dx*xi(i,j,1)+(jj-j)*dy*xi(i,j,2);
                    % relative position along disln direction
                    if abs(d_xi)<=2*dx
                        continue
                    end
                    n_count=n_count+1;
                    dVbydxi(i,j)=dVbydxi(i,j)+(V(ii,jj)-V(i,j))/d_xi;
                end
            end
        end
        if n_count>0
            dVbydxi(i,j)=dVbydxi(i,j)/n_count;
        end
        end
    end
    end
```

The above function may be called in the main integration loop as follows:

```
[~,rot]=rotation1(theta,QQ,V,dt,dx,dy,dis_width,
rho_min);
theta=theta+rot;
:
```

where $V(i,j)$ is the prescribed velocity field, dx and dy are mesh spacings, dis_width is a core width of the dislocation, and rho_min is a minimum value of density below which the density is regarded as zero.

The above approach may handle small rotations of lines, but for large rotations, the evolving $\rho(\vec{r})$ and $\theta(\vec{r})$ may lose synchronization, i.e., the $\theta(\vec{r})$ distribution no longer matches the line shape represented by the $\rho(\vec{r})$ distribution. To resume synchronization between the $\rho(\vec{r})$ and $\theta(\vec{r})$ fields, the following steps may be used:

- Curve-fit the current $\rho(\vec{r})$ distribution to a line.
- Using the fitted line from the above step, re-generate the $\theta(\vec{r})$ distribution using the method in section 3.4.1.

The above steps may be executed by the following 'redistribute' function:

```
function[QQ,theta,h_a,lx,ly,ln,xQ,yQ,ha]=redistribute(QQ,meshpoints,XX,YY,h_a,x,sdev,
trim)
% trim is a factor to preserve only high enough density for defining the line
% XX(i,j) and YY(i,j) are x- and y-coordinates of the mesh points
QQ(QQ<0)=0;
QSum0=sum(sum(QQ));
Qmax=max(QQ,[],'all','linear');
Qmin=trim*Qmax;
n = 0;
for i = size(QQ,1):-1:1
    for j = 1:size(QQ,2)
        if QQ(i,j)>=Qmin
            n = n+1;
            xQ(n)=XX(i,j);  % (xQ,yQ) are points with high enough density for
            yQ(n)=YY(i,j);  % defining the curve, indexed by n
            ha(n)=h_a(j,i); % h_a = arc length on last curve closest to (xQ,yQ)
        end
    end
end

[~,ha_sort]=sort(ha);   % ha_sort is the order of ha
xQ=xQ(ha_sort);         % xQ(n), yQ(n) now sorted according to arc length along the
yQ=yQ(ha_sort);         % old curve fitted in last cycle (since h_a is so supplied)

ln=1:1:n;
lx=csaps(ln,xQ,0.1,ln); % curve fitting using csaps function
ly=csaps(ln,yQ,0.1,ln);
line=[lx;ly];           % line is the re-fitted curve

%% coarse-graining section
theta = zeros(size(QQ,1),size(QQ,2));
[~,distance,t_a]=distance2curve(line.',meshpoints,'linear');
[QQ,theta,h_a]= discoarse2(distance,t_a,sdev,x,line,QQ,theta);
% discoarse2 is same as "discoarse" except to also return updated h_a for next cycle
QSum=sum(sum(QQ));
QQ=QQ*QSum0/QSum;
end
```

3.4.3.3 Programme for rotating arm

Re-synchronization of the $\rho(\vec{r})$ and $\theta(\vec{r})$ fields by the above 'redistribute' function may be performed once after every few steps of Runge–Kutta integration by the 'rotation1' function. The flow chart in figure 3.11 shows how this can be done (ignore the pinch-off steps marked in blue at this stage). To illustrate a situation of large rotation, rotation of a line segment about the origin of the slip plane is considered here. Note that the rotating line segment is not a dislocation since it has end points, which are not allowed for a dislocation. Figure 3.12 shows the simulation snapshots obtained using the following codes which execute the flow chart in figure 3.11. A caveat of this simulation is that the free end of the line is a source of instability due to the abrupt termination of density there; hence how long the simulation can run is quite sensitive to the preset parameters like V_max.

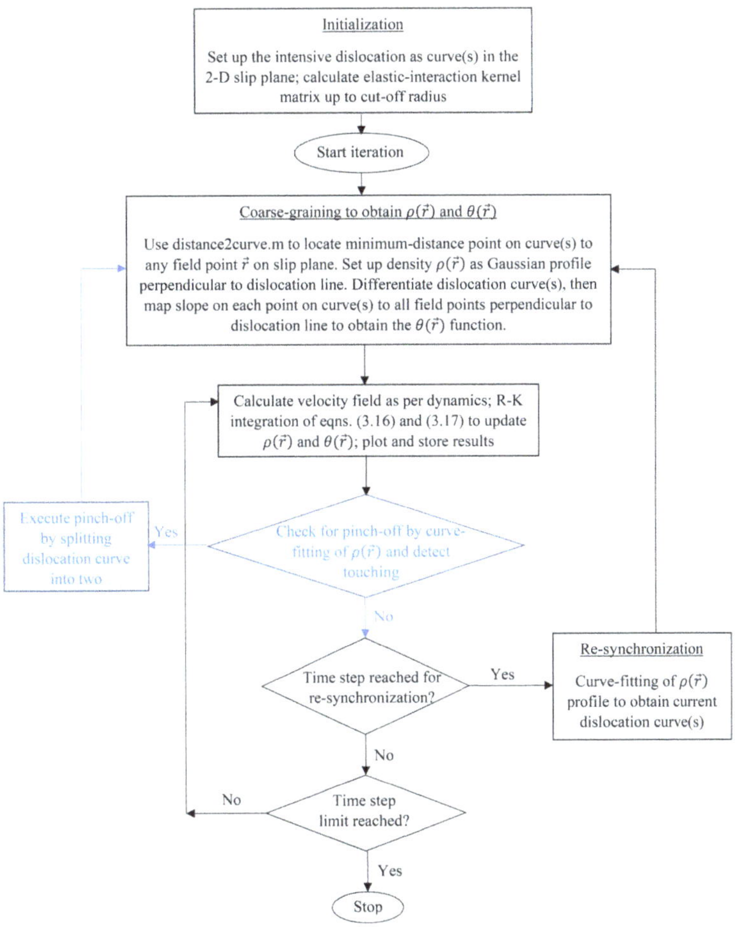

Figure 3.11. Flow chart of simulation of intensive microstructures by the GND representation. The procedures marked in blue are pertinent to microstructures involving pinch-off, such as Frank-Read source and Orowan looping. Adapted from Kalaei *et al* (2022), copyright (2022), with permission from Elsevier.

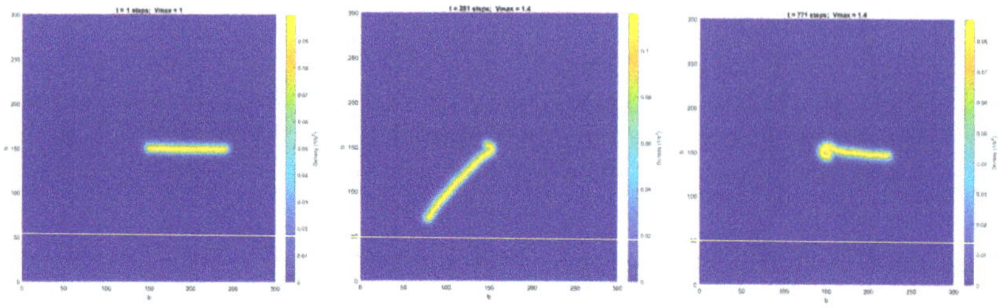

Figure 3.12. Rotating arm simulated using the rhombus mesh in section 3.4.3.1 (horizontal axis here is along *i* and vertical axis along *j*) and the movement-tilting re-synchronization method in section 3.4.3.2, using the codes in section 3.4.3.3. left = initial configuration, middle = rotated clockwise by ~135°, right = rotated after one revolution clockwise.

```
%% DDFD_Rotation_discretek: for rotating arm
%% Preset parameters
b=1;                        % Working length scale
V_max = 1.4;                % maximum velocity
rho_min=0.000001;           % minimum density to skip consideration
dis_width = 5*b;            % Width of edge dislocation core
dt= 1;                      % time step
T=1800*dt;                  % time to finish
magnitude=300*b;            % size of simulation cell
xCenter = magnitude/2;      % Center in the meshpoint coordinate system
yCenter = magnitude/2;
nxintervals = 151;          % number of x intervals 101
sdev=dis_width*0.8;         % standard deviation for gaussian profile of dislocation core

%% Set up mesh (horizontal XX axis is along i, vertical YY axis is along j)
x=linspace(0*magnitude,1*magnitude,nxintervals);
y=x;
dx=(x(end)-x(1))/(size(x,2)-1);
dy=(y(end)-y(1))/(size(y,2)-1);
meshpoints=zeros((size(x,2))^2,2);
k=1;
for i=1:size(x,2)
    for j=1:size(y,2)
        meshpoints(k,1)=x(1,i);
        meshpoints(k,2)=y(1,j);
        k=k+1;
    end
end
XX=zeros(size(x,2),size(x,2));
YY=XX;
for i=1:size(x,2)
    for j=1:size(y,2)
        XX(i,j)= x(1,i);  % XX(i,j) and YY(i,j) are another representation of the
        YY(i,j)= y(1,j);  % mesh for use by the "redistribute" function
    end
end

%% Establish dislocation line
rl= 0:dx:0.3*magnitude;
phi = 0;
lx = rl*cos(phi)+xCenter;
ly = rl*sin(phi)+yCenter;
line=[lx;ly];  % a matrix for the discrete points of the arc
```

```
%% coarse-graining section
Q=zeros(size(x,2),size(y,2)); % Q(i,j) is density over spatial domain (i,j).
QQ=Q; % QQ = 2D matrix for working density (rho), and is here initialized to Q
theta = zeros(size(x,2),size(y,2)); % theta = dislocation character (from -XX axis)
[xy,distance,t_a]=distance2curve(line.',meshpoints,'linear');
[QQ,theta,h_a]=discoarse2(distance,t_a,sdev,x,line,Q,theta); % same as "discoarse" in
section 3.4.2.2 except to also return the fractional arc length h_a

%% Presetting velocity field V(i,j)
V=zeros(size(XX,1),size(XX,2));


for i=1:size(XX,1
    for j=1:size(XX,2)
        r=sqrt((XX(i,j)-magnitude/2)^2+(YY(i,j)-magnitude/2)^2);
        V(i,j)=2*r/magnitude*V_max;  % V is proportional to radius from origin
    end
end

%% PDE solving
for time=1:round(T/dt)
    [~,rot]=rotation1(theta,QQ,V,dt,dx,dy,dis_width,rho_min);
    theta=theta+rot;
    QQ = difff2(QQ,dt,V/dx,theta);
    QQ(QQ<rho_min)=0;
    if mod(time,4)==1         % Perform synchronization once every 4 time steps
       [QQ,theta,h_a,lx,ly,ln,xQ,yQ,ha]=
                                      redistribute(QQ,meshpoints,XX,YY,h_a,x,sdev,0.9);
    end
    %
    % Density QQ can be plotted here, which is skipped for clarity purpose
    %
end
```

References

D'Errico J 2010 distance2curve. https://mathworks.com/matlabcentral/fileexchange/34869-distance2curve *MATLAB Central File Exchange*. Released 2010

Hochrainer T, Zaiser M and Gumbsch. P 2007 A three-dimensional continuum theory of dislocation systems: kinematics and mean-field formulation *Phil. Mag.* **87** 1261

Kalaei A, Xiang Y and Ngan. A H W 2022 An efficient and minimalist scheme for continuous dislocation dynamics *Int. J. Plast.* **158** 103433

Sedláček R, Kratochvíl J and Werner E 2003 The importance of being curved: bowing dislocations in a continuum description *Phil. Mag.* **83** 3735–52

Walgraef D and Aifantis E C 1985 Dislocation patterning in fatigued metals as a result of dynamical instabilities *J. Appl. Phys.* **58** 688–91

Xia S and El-Azab A 2015 Computational modelling of mesoscale dislocation patterning and plastic deformation of single crystals *Model. Simul. Mater. Sci. Eng.* **23** 055009

IOP Publishing

Continuous Dislocation Dynamics Made Simple

Alfonso Ngan and Alireza Kalaei

Chapter 4

Dynamics of intensive line density

4.1 Dynamics closure by velocity law

In chapter 3, we have seen the various forms of kinematics law for intensive dislocations, and simulation examples of expanding dislocation loop and rotating arm. In the case of the expanding dislocation loop, the velocity field $v(\vec{r})$ was set to be a constant field while for the rotating arm, $v(\vec{r})$ was set to be proportional to the radial distance $\vec{r}$ from the centre of rotation. For real dislocations, as mentioned in chapter 2, the velocity field is governed by a dynamics law which will close the problem.

Traditionally, etch-pit experiments are conducted to measure dislocation velocity under an applied stress. In these experiments, a dislocation-containing crystal is subjected to chemical etching after surface polishing to form etch-pits preferentially at locations where dislocations emerge on the free surface. The etch-pits therefore mark the positions of the dislocations in the crystal. Then, etching and observation before and after stress application would reveal how far and fast the dislocations have travelled under stress. Figure 4.1(a) shows typical results of such experiments. Alternatively, MD can also be carried out to study the speed of dislocations under different stress, as illustrated by the example in figure 4.1(b). From plenty of such data gathered in the past, there is a general consensus that dislocation velocity v can be described by a law of the form $v \propto \tau^n$, where τ is the applied stress. For quite a wide range of velocity (see figure 4.1), the relation is proportional, i.e., $n = 1$. Therefore, in DDD simulations, the velocity law is usually assumed as a proportional relation with the effective resolved shear stress $\tau^{\text{eff}}(\vec{r})$ on the slip system:

$$v(\vec{r}, \theta) = m(\theta)\, \tau^{\text{eff}}(\vec{r}) \tag{4.1}$$

where $m(\theta)$ is the dislocation mobility, which in general depends on the dislocation character θ.

The effective resolved shear stress is generally considered to comprise two components:

doi:10.1088/978-0-7503-6330-3ch4

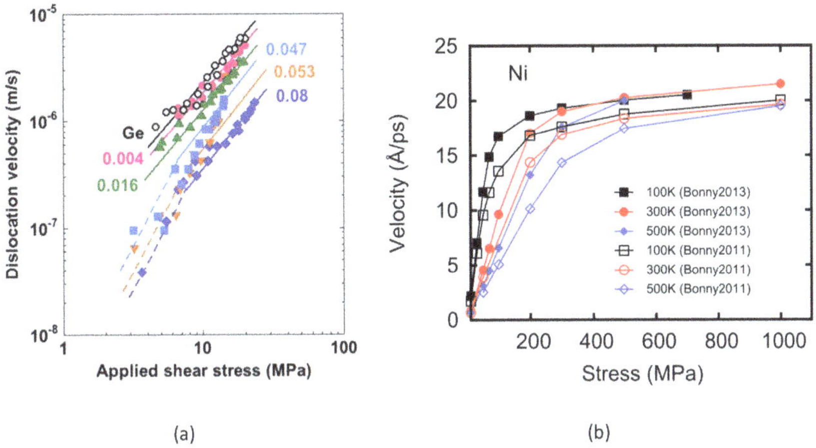

Figure 4.1. Dislocation velocity laws. (a) Experimentally determined laws from etch-pit experiments in Si-Ge alloys with different Si concentrations (reproduced with permission from Yonenaga (2013) CC BY 3.0. (b) MD calculated dislocation velocity laws in Ni using different potentials (reproduced from Zhao *et al* (2017), copyright (2017) with permission from Elsevier).

$$\tau^{\text{eff}}(\vec{r}) = \tau^{\text{int}}(\vec{r}) + \tau^{\text{app}} \tag{4.2}$$

where τ^{app} is the externally applied resolved shear stress, and $\tau^{\text{int}}(\vec{r})$ is the internal interaction stress. This chapter aims to address the $\tau^{\text{int}}(\vec{r})$, and to present examples of intensive dislocation structures simulated with dynamics closure.

For an intensive dislocation microstructure, the $\tau^{\text{int}}(\vec{r})$ term arises from two contributions, namely,

(i) the long-range elastic interactions between dislocation segments present in the system, and
(ii) interactions within the dislocation core akin to those considered in the Peierls–Nabarro model.

4.2 Mura's formula for long-range elastic interaction

4.2.1 Mura's formula for strain field of a dislocation loop

It is well-known that dislocations are strain centres causing distortions in the lattice. In the literature, different forms of dislocation strain fields exist but T Mura in 1963 derived a very useful form for our present purpose, namely, the elastic displacement gradient (i.e. strain) field $\partial u_r/\partial x_s$ of an arbitrary dislocation loop C in space, which is given by a line integral along C:

$$\frac{\partial u_r}{\partial x_s}(\vec{r}) = \epsilon_{jsh} b_i c_{ijkl} \oint_c \frac{\partial G_{rk}(\vec{R})}{\partial x_l{}'} dx_h{}' \tag{4.3}$$

Here, $\vec{r} = (x_1, x_2, x_3)$ is the 'field' point at which the strain is to be calculated, $\vec{r}' = (x_1', x_2', x_3')$ is a 'source' point along the dislocation loop C, and $G_{rk}(\vec{R})$ is the elastic Green's function with $\vec{R} = \vec{r}' - \vec{r}$. ϵ_{jsh} is the Einstein permutation operator (also known as the Levi-Civita symbol), which is equal to 1 when *jsh* are in cyclic order, −1 when they are in reverse cyclic order, and 0 otherwise.

Why is this form so relevant to us? Since the line integration in equation (4.3) is simply a summation of infinitesimal line segments along the dislocation loop, Mura's formula allows the elastic strain at any field point in the system to be calculated as a summation of the effects of dislocation contents at different source points in the system, which tallies well with our approach of (i) describing dislocation contents as densities, and (ii) computational simulation where summation can be conveniently carried out by a suitable numerical scheme.

Different derivations of Mura's formula are available in the literature, and here we take the approach by Hirth and Lothe (1992), which may be the easiest to understand. To start with, let us first examine the Green's function in equation (4.3). $G_{rk}(\vec{R})$ of an elastic medium is the displacement along direction k at point $\vec{R}$ due to a point force along direction r exerted at the origin (k, r = 1–3). Therefore, for a point force $\vec{f}(\vec{r})$ acting at point $\vec{r}$ in an infinite elastic medium, with components f_r (r= 1–3) along the three Cartesian axes of space, the displacement field due to $\vec{f}(\vec{r})$ is given by:

$$\tilde{u}_k(\vec{r}\,') = G_{rk}(\vec{R})f_r(\vec{r}) \tag{4.4}$$

where $\vec{R} = \vec{r}\,' - \vec{r}$ is the position where displacement $\vec{u}$ is measured, relative to position where the point force $\vec{f}$ acts. (The reader may notice that $\vec{f}$ acts on $\vec{r}$, which was said to be the field point above, and $\tilde{u}_k$ is measured at $\vec{r}\,'$, which was said to be the source point. As will be clear if one reads on, the source and field points here refer to the eventual strain field of the dislocation, rather than the Green's function in equation (4.4).) The strain field is then the derivative of $\tilde{u}_k(\vec{r}\,')$ with respect to movements dx_l' in $\vec{r}\,'$:

$$\frac{\partial \tilde{u}_k}{\partial x_l'}(\vec{r}\,') = \frac{\partial G_{rk}}{\partial x_l'}(\vec{R})f_r(\vec{r}) \tag{4.5}$$

From Hooke's law, the stress field of the point force is given as:

$$\tilde{\sigma}_{ij}(\vec{r}\,') = c_{ijkl}\frac{\partial \tilde{u}_k}{\partial x_l'}(\vec{r}\,') = c_{ijkl}\frac{\partial G_{rk}}{\partial x_l'}(\vec{R})f_r(\vec{r}) \tag{4.6}$$

Next, let us create a dislocation loop C in the vicinity of the point force $\vec{f}$ at $\vec{r}$. The reader will recall the following procedure to create a dislocation: First, the elastic medium is cut open along a surface A bounded by the dislocation curve C. Then, every material just above A is displaced relative to the adjoining material just below A by the Burgers vector $\vec{b}$, and then the cut surface is rewelded, with any overlapping material removed or gap filled if necessary. As shown in figure 4.2, the cut surface A does not need to be flat and the curve C does not need to be co-planar. Therefore, in this process of dislocation creation, displacements will occur in the elastic medium, and the work done by $\vec{f}$ on the elastic medium as the dislocation is created in the latter is:

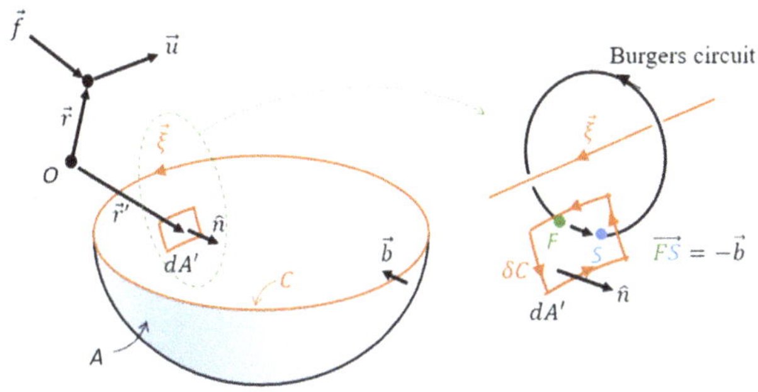

Figure 4.2. Creation of a dislocation loop C in the vicinity of a point force $\vec{f}$ at $\vec{r}$. Relative displacement by $\vec{b}$ (Burgers vector) is made across any elemental area dA' in surface A bounded by C. With the convention for $\vec{b}$ in figure 1.2, for a broken RH Burgers circuit drawn on the dislocated crystal, the vector joining the finish (F) to the start (S) is $-\vec{b}$; hence the material point (in blue) just above dA' in the positive normal direction is displaced by $-\vec{b}$ relative to the material point (in green) just below dA'.

$$W = -\iint_A b_i\, \tilde{\sigma}_{ij}(\vec{r}\,')dA_j' \tag{4.7}$$

Here, dA_j' is the j-axis projection of an infinitesimal element of A at $\vec{r}\,'$, $\tilde{\sigma}_{ij}(\vec{r}\,')dA_j'$ is the force component along i at $\vec{r}\,'$, due to the stress field of the point force $\vec{f}$ at $\vec{r}$, and hence $-b_i\tilde{\sigma}_{ij}(\vec{r}\,')dA_j'$ (summed over i and j) is the work done at point $\vec{r}\,'$ on the cut surface A. The minus sign before b_i here follows the finish-start/right-hand (FS/RH) convention for Burgers vector definition—with the dislocation line direction $\hat{\xi}$ along C as shown in figure 4.2, the normal direction $\hat{n}$ of an elemental area dA' will be pointing inward the curved surface A, and with a Burgers circuit in the RH sense about $\hat{\xi}$, the material on the 'above' (i.e. along $+\hat{n}$) side of dA' will be displaced by $-\vec{b}$ relative to the material on the 'below' (i.e., along $-\hat{n}$) side of dA', hence the displacement on dA' is $-\vec{b}$ instead of $+\vec{b}$.

The W in equation (4.7) is the interaction energy between the dislocation loop and the point force, i.e., it is the difference in the formation energy of the dislocation with and without the presence of the point force. Substituting in equation (4.6) for the $\tilde{\sigma}_{ij}(\vec{r}\,')$, the interaction energy becomes:

$$W = -f_r(\vec{r})\, b_i c_{ijkl} \iint_A \frac{\partial G_{rk}}{\partial x_l'}(\vec{r}\,' - \vec{r})\, dA_j' \tag{4.8}$$

The created dislocation will introduce a displacement field $u_r(\vec{r})$ in the medium, and so the same interaction energy in equation (4.8) can also be viewed as the product of $u_r(\vec{r})$ and $f_r(\vec{r})$ summed over $r = 1$ to 3, i.e.,

$$W = f_r(\vec{r})\, u_r(\vec{r}) = -f_r(\vec{r})\, b_i c_{ijkl} \iint_A \frac{\partial G_{rk}}{\partial x_l'}(\vec{r}\,' - \vec{r})\, dA_j' \tag{4.9}$$

Since the components f_r and location of force application $\vec{r}$ are arbitrary, to make equation (4.9) hold for all cases, the displacement field of the dislocation loop must be given as:

$$u_r(\vec{r}) = -b_i c_{ijkl} \iint_A \frac{\partial G_{rk}}{\partial x_l'}(\vec{r}' - \vec{r})\, dA_j' \tag{4.10}$$

which was first derived by Volterra in 1907. The strain field of the dislocation loop can be obtained by differentiating equation (4.10) with respect to movements dx_s in the field point $\vec{r}$:

$$\frac{\partial u_r}{\partial x_s}(\vec{r}) = -b_i c_{ijkl} \iint_A \frac{\partial^2 G_{rk}}{\partial x_s \partial x_l'}(\vec{r}' - \vec{r})\, dA_j' = b_i c_{ijkl} \iint_A \frac{\partial^2 G_{rk}}{\partial x_s' \partial x_l'}(\vec{r}' - \vec{r})\, dA_j' \tag{4.11}$$

The second step above involves a change from ∂x_s to $-\partial x_s'$. This is valid because the Green's function G_{rk} and its first derivatives $\partial G_{rk}/\partial x_l'$ depend only on the relative vector $\vec{R} = \vec{r}' - \vec{r}$, and the same change in $\vec{R}$ can be brought by an increase in x_s' or a decrease in x_s of the same magnitude, hence $\partial x_s = -\partial x_s'$ which explains the second step.

To proceed to the Mura formula, we note that the condition of mechanical equilibrium for the stress field $\tilde{\sigma}_{ij}$ of the point force is:

$$\frac{\partial \tilde{\sigma}_{ij}(\vec{r}')}{\partial x_j'} = 0 \tag{4.12}$$

which, in conjunction with equation (4.6), implies that

$$c_{ijkl} \frac{\partial^2 G_{rk}}{\partial x_j' \partial x_l'} = 0 \tag{4.13}$$

A null term, $-b_i \iint_A c_{ijkl} \frac{\partial^2 G_{rk}}{\partial x_j' \partial x_l'} dA_s'$, can therefore be added to the right side of equation (4.11) without affecting its validity, i.e.,

$$\frac{\partial u_r}{\partial x_s}(\vec{r}) = b_i c_{ijkl} \iint_A \left[\frac{\partial^2 G_{rk}}{\partial x_s' \partial x_l'} dA_j' - \frac{\partial^2 G_{rk}}{\partial x_j' \partial x_l'} dA_s' \right] = \epsilon_{jsh} b_i c_{ijkl} \oint_c \frac{\partial G_{rk}(\vec{r}' - \vec{r})}{\partial x_l'} dx_h' \tag{4.14}$$

where Stoke's theorem is applied in the second step. The result is the Mura formula in equation (4.3).

4.2.2 Mura's formula for continuous distribution of dislocations

Equation (4.3) is for an arbitrary dislocation loop in space, but Mura also generalized it for a continuous distribution of dislocations described by the Nye tensor (Mura 1987). Here, we show how this generalization can be done, using the dislocation density that we have defined in chapter 2 (equations (2.8) and (2.9)). Consider a general dislocation–density picture where different slip systems (labelled by α) co-exist, and $\rho_{\alpha'}(\vec{r}')$ is the dislocation density (defined in equation (2.9)) for the

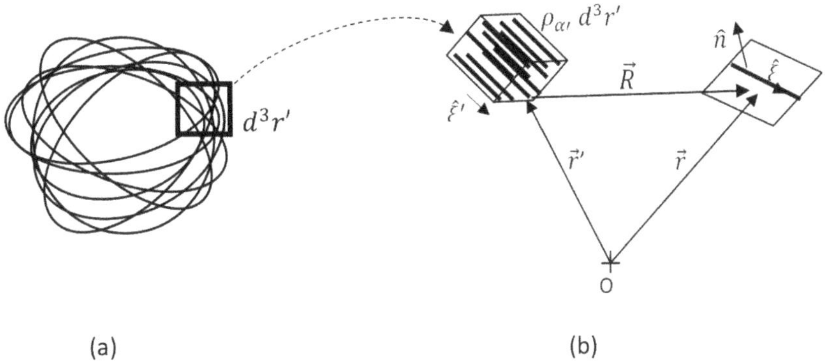

Figure 4.3. (a) A system of a large number of dislocation loops on slip system α' can be converted into a continuous density distribution $\rho_{\alpha'}(\vec{r}')$ where $\rho_{\alpha'}\, d^3r'$ gives the total line length within a small pixel d^3r' at $\vec{r}'$ due to the loops passing through there. (b) Dislocation line lengths $\rho_{\alpha'}\, d^3r'$ within small pixels d^3r' at different source points $\vec{r}'$ then contribute to the displacement-gradients (i.e. strains) at a given field point $\vec{r}$.

slip system α' at point $\vec{r}'$. The displacement gradient at a general field point $\vec{r} = (x_1, x_2, x_3)$ is due to the elastic stress fields of all the dislocations present in the system, each of which may be assumed to be a closed loop of the type considered in equation (4.3), since dislocations do not terminate inside a continuum. Then, in order to obtain the strain at a given field point $\vec{r}$, summing up the line integrals of all the loops in space, each given by equation (4.3), would be the same as summing up the effects of the dislocation contents present at different source points $\vec{r}'$ in the system (figure 4.3), by the principle of elastic superposition. Therefore, the strain field can be obtained by making the substitution $dx_h' = \xi_h^{\alpha'}\, \rho_{\alpha'}(\vec{r}')\, d^3r'$ in equation (4.3), followed by integrating over all source points $\vec{r}' = (x_1', x_2', x_3')$ and summing over all slip systems α':

$$\frac{\partial u_r}{\partial x_s}(\vec{r}) = \epsilon_{jsh} b_i c_{ijkl} \oint_c \frac{\partial G_{rk}(\vec{R})}{\partial x_l'} dx_h' = \sum_{\alpha'} \epsilon_{jsh}\, b_i^{\alpha'}\, c_{ijkl} \iiint \frac{\partial G_{rk}(\vec{R})}{\partial x_l'} \xi_h^{\alpha'}\, \rho_{\alpha'}(\vec{r}')\, d^3r' \quad (4.15)$$

Here, $\xi_h^{\alpha'}\, \rho_{\alpha'}(\vec{r}')$ is the projection of the dislocation–density vector $\vec{\rho}_{\alpha'}(\vec{r}')$ (see equation (2.8)) onto the infinitesimal line segment $\{dx_h'\}$ along curve C.

τ^{int} on slip system α is related to the displacement gradients (i.e. strains) by Hooke's law:

$$\tau_\alpha^{\text{int}}(\vec{r}) = n_p^\alpha\, b_q^\alpha\, c_{pqrs} \frac{\partial u_r}{\partial x_s}(\vec{r}) \quad (4.16)$$

Substituting in equation (4.15), we have

$$\tau_\alpha^{\text{int}}(\vec{r}) = n_p^\alpha\, b_q^\alpha\, c_{pqrs} \sum_{\alpha'} \epsilon_{jsh}\, b_i^{\alpha'}\, c_{ijkl} \iiint \frac{\partial G_{rk}(R)}{\partial x_l'} \xi_h^{\alpha'}\, \rho_{\alpha'}(\vec{r}')\, d^3r' \quad (4.17)$$

which is Mura's generalized formula for a continuous distribution of dislocations (Mura 1987, p 340, equation (37.13)), now cast in terms of dislocation densities $\rho_{\alpha'}(\vec{r}')$ on different slip systems α'.

In the rest of this book, we shall limit our discussion to dislocation problems in infinite, isotropically elastic media, for which the elastic Green's function is given in chapter 2 of Hirth and Lothe (1992):

$$G_{rk}(R) = \frac{1}{8\pi\mu}\left[\delta_{rk}\,\nabla^2 R - \frac{1}{2(1-\nu)}\frac{\partial^2 R}{\partial x_r'\partial x_k'}\right] \tag{4.18}$$

where μ is shear modulus, ν is Poisson's ratio, and δ_{rk} is the Kronecker delta (equals 1 when $r = k$ and zero otherwise). With $R = |\vec{r}' - \vec{r}|$, $\nabla^2 R$ can be worked out to be $2/R$ and similarly, the second term above would comprise terms of $1/R$ and $1/R^3$. The reader will immediately see a problem, namely, if the ordinary definition of $R = |\vec{r} - \vec{r}'|$ in figure 4.3(b) is used, then when $\vec{r}' = \vec{r}$, $R = 0$ and so $G_{rk}(R)$ will be singular, which will pose problems in numerical computation. The same problem also exists in DDD simulation, and to overcome this, Cai *et al* (2007) proposed to add a small, atomic length scale a to the R, so that

$$R = \sqrt{|\vec{r} - \vec{r}'|^2 + a^2} = \sqrt{\sum\nolimits_{i=1}^{3}\left(x_i - x_i'\right)^2 + a^2} \tag{4.19}$$

Their argument is that dislocations are present in a lattice rather than in a continuum, and so the situation of the field point (where strain is measured) overlapping with the source point (where dislocation content resides) should never happen; at most, they should be separated by one atomic bond apart, hence the length a. Therefore, if the field and source points are far apart, which will be the case for considering long-range interaction between dislocation segments, then $|\vec{r} - \vec{r}'| \gg a$ and R will approach the old definition of $|\vec{r} - \vec{r}'|$. But when we are considering interactions within a dislocation core, then we simply treat a source point overlapping with the field point as slightly displaced by $R = a$, which is embodied in equation (4.19).

Therefore, in what follows, we will use Cai *et al*'s definition in equation (4.19) for R to remove the mathematical singularity in $G_{rk}(R)$. With R defined in equation (4.19), carrying out the differentiation in equation (4.18) will give the following explicit expression for $G_{rk}(R)$:

$$G_{rk}(R) = \frac{1}{16\pi\mu(1-\nu)}\left[\frac{(3-4\nu)\delta_{rk}}{R} + \frac{\left(x_r - x_r'\right)\left(x_k - x_k'\right)}{R^3} + \frac{2a^2(1-\nu)\delta_{rk}}{R^3}\right] \tag{4.20}$$

Then, the differentiation $\frac{\partial G_{rk}(R)}{\partial x_l'}$ in equation (4.17) can be carried out term by term to obtain the following form of the long-range elastic stress field due to a dislocation–density distribution:

$$\tau_\alpha^{\text{int}}(\vec{r}) = \sum\nolimits_{\alpha'}\iiint g_{\alpha\alpha' h}(\vec{r} - \vec{r}')\xi_h^{\alpha'}\rho_{\alpha'}(\vec{r}')d^3r' \tag{4.21A}$$

where the kernel is

$$g_{\alpha\alpha' h}(\vec{r} - \vec{r}') = \frac{n_p^{\alpha} b_q^{\alpha} c_{pqrs}}{16\mu\pi(1 - \nu)} \epsilon_{jsh} b_i^{\alpha'} c_{ijkl} \times \left[\frac{6\delta_{rk} a^2 (1 - \nu)\left(x_l - x_l'\right)}{R^5} + \frac{(3 - 4\nu)\delta_{rk}\left(x_l - x_l'\right)}{R^3} + \frac{3\left(x_r - x_r'\right)\left(x_k - x_k'\right)\left(x_l - x_l'\right)}{R^5} - \frac{\delta_{rl}\left(x_k - x_k'\right) + \delta_{kl}\left(x_r - x_r'\right)}{R^3} \right] \tag{4.21B}$$

Furthermore, for isotropic elasticity, the elastic constants are given by

$$c_{ijkl} = \mu(\delta_{ik}\delta_{jl} + \delta_{il}\delta_{jk}) + \lambda\delta_{ij}\delta_{kl}, \tag{4.22}$$

where $\lambda = \frac{2\mu\nu}{1 - 2\nu}$ is Lame's constant.

If there is only one slip system present in the system, then all α' are the same and can be dropped, and the triple integration in equation (4.21A) can be replaced by a double integration over the slip plane.

Exercise 4.1. Prove equations (4.20) and (4.21).

Solution: Let us first deal with the $\nabla^2 R$ in equation (4.18):

$$\nabla^2 R = \sum_l \frac{\partial}{\partial x_l'}\left(\frac{\partial}{\partial x_l'}\sqrt{\sum_r \left(x_r' - x_r\right)^2 + a^2}\right) = \sum_l \frac{\partial}{\partial x_l'}\left(\frac{\left(x_l' - x_l\right)}{R}\right)$$

$$= \sum_l \left[\frac{1}{R} - \frac{\left(x_l' - x_l\right)^2}{R^3}\right] = \frac{3}{R} - \frac{R^2 - a^2}{R^3} = \frac{2}{R} + \frac{a^2}{R^3}$$

Next, the second derivative in the second term in equation (4.18) is:

$$\frac{\partial^2 R}{\partial x_r' \partial x_k'} = \frac{\partial}{\partial x_r'}\left[\frac{\left(x_k' - x_k\right)}{R}\right] = \frac{\delta_{rk}}{R} - \frac{\left(x_r' - x_r\right)\left(x_k' - x_k\right)}{R^3}$$

Substituting these two terms into equation (4.18) will give equation (4.20).

Then, with equation (4.20), the $\partial G_{rk}/\partial x_l'$ in equation (4.17) can be worked out as:

$$\frac{\partial G_{rk}}{\partial x_l'} = \frac{1}{16\pi\mu(1 - \nu)}\frac{\partial}{\partial x_l'}\left[\frac{(3 - 4\nu)\delta_{rk}}{R} + \frac{\left(x_r - x_r'\right)\left(x_k - x_k'\right)}{R^3} + \frac{2a^2(1 - \nu)\delta_{rk}}{R^3}\right]$$

$$= \frac{1}{16\pi\mu(1 - \nu)}\left[\frac{(3 - 4\nu)\delta_{rk}\left(x_l - x_l'\right)}{R^3} + \frac{\delta_{kl}\left(x_r' - x_r\right) + \delta_{lr}\left(x_k' - x_k\right)}{R^3} + \frac{3\left(x_r - x_r'\right)\left(x_k - x_k'\right)\left(x_l - x_l'\right)}{R^5} - \frac{6a^2(1 - \nu)\delta_{rk}\left(x_l' - x_l\right)}{R^5}\right]$$

Substituting into equation (4.17) will lead to equations (4.21a) and (4.21b).

4.3 Interaction within dislocation core

For an intensive dislocation, as discussed in sections 2.2.2 and 2.4 in chapter 2, core interactions need to be considered. As shown in figure 2.4, the lattice disregistry $\vec{u}(\vec{r})$ inside a dislocation core has a high gradient corresponding to a sharp dislocation–

density profile, which, from the above discussion on mutual-density elastic interaction, has a strong tendency to spread. The spreading, however, will increase the area on the slip plane where lattice misfit (or atomic bond breakage) across the slip plane will occur. The lattice misfit is characterized by a specific energy per unit misfit area $\gamma(\vec{u})$, and hence, according to the Peierls–Nabarro (P–N) model, the balance between the elastic interaction energy and the lattice misfit energy determines the degree of spreading of the dislocation–density profile of the intensive dislocation. Say, for a screw or edge dislocation, the equilibrium density profile $\rho(x)$, where x is coordinate on the slip plane perpendicular to the dislocation line, is the solution to the following coupled integral equations:

$$u(x) = \int_{-\infty}^{x} \rho(x')dx'; \quad A\int_{-\infty}^{+\infty} \frac{\rho(x')}{x - x'}dx' = -\frac{d\gamma}{du} \tag{4.23}$$

where $A = \mu b/[2\pi(1 - \nu)]$ for edge dislocation, and $A = \mu b/(2\pi)$ for screw dislocation.

The ease of solving equation (4.23) would depend on the 'γ-surface' $\gamma(u)$, which is a material property, and in general, a numerical solution of equation (4.23) can be very expensive computationally. Accordingly, Zhang and Ngan (2018) demonstrated a short-cut solution to the problem, by treating the lattice misfit force as a type of force that shields the elastic repulsion among the dislocation density. The shielding effect is represented by a simple factor f incorporated into the total interaction stress in equation (4.21) as follows:

$$\tau_\alpha^{\text{int}}(\vec{r}) = \sum_{\alpha'} \iiint g_{\alpha\alpha' h}(\vec{r} - \vec{r}')\xi_h^{\alpha'} \rho_{\alpha'}(\vec{r}') \left[1 - f(d(\vec{r}, \vec{r}'))\right] d^3r' \tag{4.24}$$

Here, $f(d)$ is a core-shielding function of the form:

$$f(d) = (d/w)^n \tag{4.25}$$

where w is the desired dislocation width, n is a constant, and d is the spacing between the source point $\vec{r}'$ and field point $\vec{r}$ along the dislocation width direction, i.e.

$$d = (\vec{r} - \vec{r}') \cdot (\hat{n} \times \hat{\xi}), \tag{4.26}$$

$\hat{n}$ is slip plane normal and $\hat{\xi}$ is dislocation line direction. The effect of the $f(d)$ function in equations (4.24) and (4.25) can be easily understood from figure 4.4. If the source-field point spacing d is wider than the desired core width w, then f from equation (4.25) will be larger than 1 and then τ_α^{int} from equation (4.24) will be negative; hence an attractive force will exist to pull the densities at $\vec{r}$ and $\vec{r}'$ together. On the contrary if $\vec{r}$ and $\vec{r}'$ are closer together than w, then their densities will repel. Use of the core-shielding factor f should therefore stabilize the dislocation density profile to the desired width w. The model parameters w and n in equation (4.25) may be set to depend on the dislocation character θ, to cater for the fact that the dislocation width is dependent on θ.

The core-shielding approach in equation (4.24) is meant to be a short-cut to an otherwise expensive problem, but the result is miraculously satisfactory. Figure 4.5

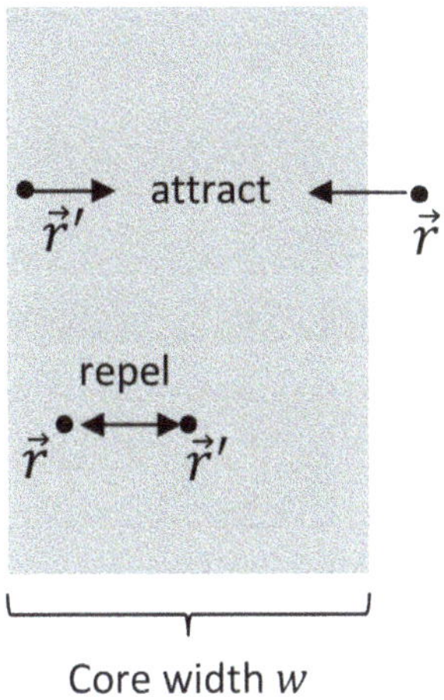

Figure 4.4. Core shielding in intensive dislocations.

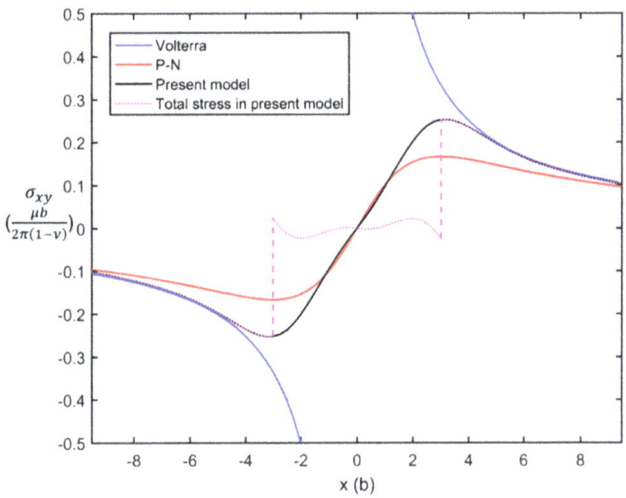

Figure 4.5. Comparison between Volterra, Peierls–Nabarro (P–N) and core-shielding models, for the computed shear stress on slip plane for edge dislocation. Reprinted from Zhang and Ngan (2018), copyright (2018), with permission from Elsevier.

compares the computed shear stress on the slip plane for an edge dislocation, between the analytical Volterra model (by treating the core as a singularity), the original, analytical P–N model [for a sinusoidal $\gamma(u)$], and the core-shielding model by substituting the analytical density distribution of the P–N model into equation (4.24). The agreement with the Volterra model outside the core is good, and inside the core the singularity is removed as for the case of the P–N model. Furthermore, the total stress inside the core computed from equation (4.24) is close to zero, indicating near equilibrium there.

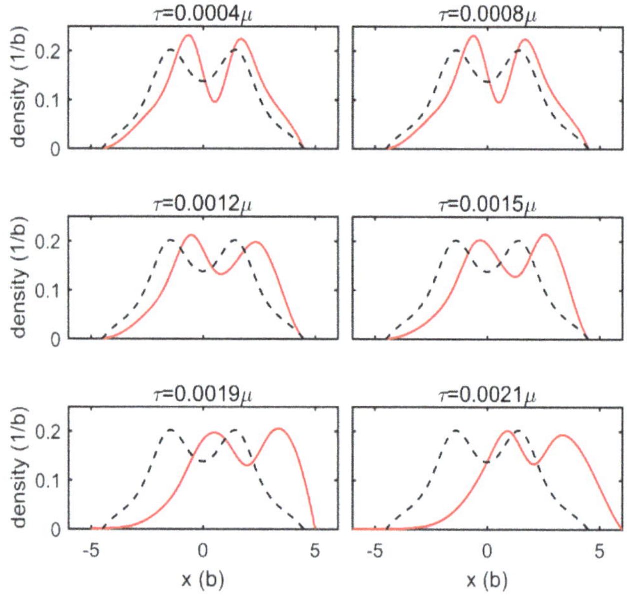

Figure 4.6. Equilibrium density distribution of an edge dislocation predicted from the core-shielding model, at different applied stress before equilibrium is lost (red lines). Black lines represent the stress-free profile. Reprinted from Zhang and Ngan (2018), copyright (2018) with permission from Elsevier.

The P–N model not only provides physical understanding on why an equilibrium core width exists, but also a resistance to dislocation motion known as the Peierls stress. However, the original P–N model predicts a Peierls stress only when the calculation of the misfit energy is carried out by discrete summation instead of continuous integration (Hirth and Lothe 1992, chapter 8, pp 236–7, equations (8.37) and (8.38)), i.e., it is more of a mathematical consequence than physical. In the present core-shielding model, due to the stabilizing role of the stress-shielding factor, as a mild stress is applied the density profile simply changes shape without losing equilibrium, as shown in figure 4.6. Equilibrium is lost and hence the dislocation begins to move only when the applied stress is large enough, i.e., the core-shielding model also predicts a Peierls stress. The Peierls stress, density profile and γ-surface $\gamma(u)$ can also be tuned by the model parameters n and w in equation (4.25), as shown in figures 4.7 and 4.8. In figure 4.8, $\gamma(u)$ is obtained from the P–N model by integrating the second equation in equation (4.23) after substituting in the $\rho(x)$ profile calculated from the core-shielding model. All these amount to say that the core-shielding model connects well with conventional understanding of dislocation core physics, in terms of Peierls stress and γ-surface.

4.4 Simulation codes for internal interaction stress τ^{int}

4.4.1 Mura's elastic interaction

Here, we discuss the implementation of the long-range elastic interaction between dislocation densities. We consider the simpler case of a single slip system for a single

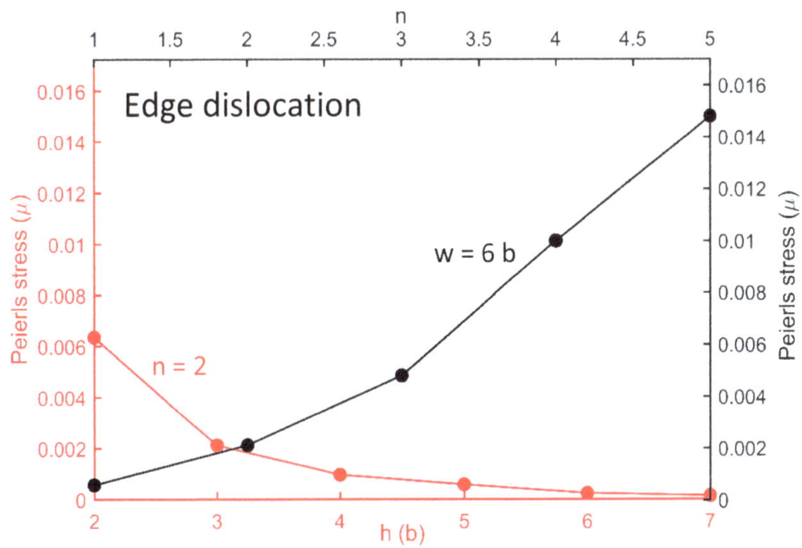

Figure 4.7. Effects of model parameters (n and w) on the Peierls stress predicted from the core-shielding model. In the lower abscissa, $h = w/2$ is the half-width of the dislocation. (Adapted from Zhang and Ngan (2018), copyright (2018), with permission from Elsevier).

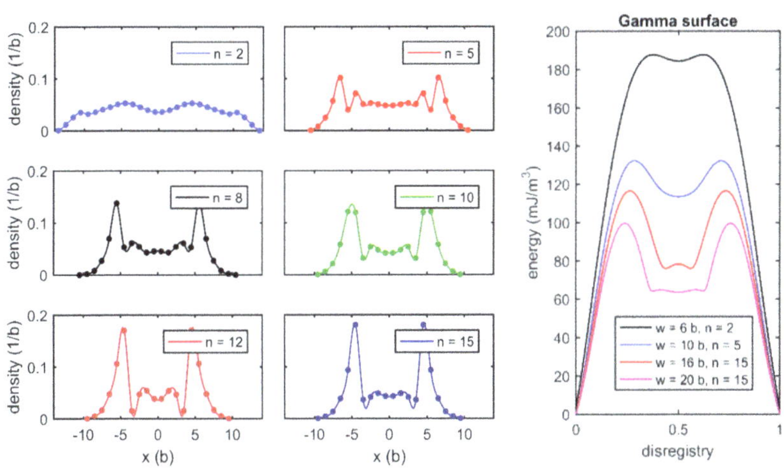

Figure 4.8. Effects of model parameters (n and w) on the density distribution (left) and γ surface (right). For the density distributions, $w = 20b$. (Reproduced from Zhang and Ngan (2018), copyright (2018) with permission from Elsevier).

slip plane. Therefore, in equation (4.24), $\alpha' = \alpha$, the dislocation density $\rho_\alpha(\vec{r}')$ is defined only on the 2D slip plane so that the integration is only 2D over the slip plane, and the line orientation ξ_h^α has only two components (i.e., $h = 1$ or 2 only).

The following function interaction_coeff will return an array int_c_lr(*ii*,*jj*,*h*), which is the kernel $g_{\alpha\alpha h}(\vec{R}(ii, jj))$ in equation (4.21B), where $\vec{R}(ii, jj) = -\vec{r}'_{(ii, jj)}$ is minus of the position vector of the source point (ii,jj) from the 'origin', up to a pre-set cut-off radius Rc_int. When the int_c_lr(ii,jj,h)'s are used later (by function interaction_lr_core1 discussed in the next subsection), the 'origin' here will be shifted to each field point $\vec{r}$ at which the interaction stress τ_α^{int} (see equation (4.24)) is calculated. In equation (4.24), ξ_h^α is the projection of the dislocation-direction unit vector onto the screw ($h = 1$, XX) or edge ($h = 2$, YY) direction.

```
function [int_c_lr]=interaction_coeff(n_vec, b_vec, a, mu, nu, dx, dy, Rc_int)

% Long-range interaction coefficients in Mura's formulation for interaction stress
% simplified for single slip system, i.e. alpha=alpha'
% int_c_lr(ii,jj,h) = n_p*b_q*C_pqrs/mu*C_klij*b_i*e_sjh*G_rk,l for each (x-x')
vector specified by (ii,jj,0) on slip plane, c.f. eqn. (4.17).
% within cutoff distance, with h specifying the source-dislocation line direction
(i.e. d_l(h)).
% a is atomic size that removes singularity in Mura's formula

lambda=2*mu*nu/(1-2*nu);
lom=lambda/mu;
nx_co=ceil(Rc_int/dx); % Rc_int = cutoff radius of dislocation interaction
ny_co=ceil(Rc_int/dy);
int_c_lr=zeros(nx_co*2+1,ny_co*2+1,2);
[e]=epsi; % permutation tensor

% Elastic constants
% C_ijkl = mu*(delta_ik*delta_jl+delta_il*delta_kj) + lambda*(delta_ij*delta_kl)
cbymu = zeros(3,3,3,3);
for i = 1:3
    for j = 1:3
        for k = 1:3
            for l = 1:3
                cbymu(i,j,k,l) = (i==k)*(j==l)+(i==l)*(k==j)+lom*(i==j)*(k==l);
            end
        end
    end
end

%Long range interaction coefficients; eqns. 4.17 & 4.21b
for ii=1:nx_co*2+1
    for jj=1:ny_co*2+1
        x=-[dx*(ii-nx_co-1) dy*(jj-ny_co-1) 0];
% x(3) is the (x-x') vector on the slip plane within cutoff,
% each (x-x') is specified by (ii,jj) on slip plane within cutoff;
% int_c_lr(ii,jj,h) = n_p*b_q*C_pqrs/mu*C_klij*b_i*e_sjh*G_rk,l
% for each (x-x') within cutoff specified by (ii,jj).
        R=sqrt(dot(x,x));
        R_revo=(R^2+(a*dx)^2)^0.5;   % eq. 4.19 with Cai et al.'s singularity removal
        for j=1:3
            for s=1:3
                for r=1:3
                    for p=1:3
                        for q=1:2
                            % t1 = n_p * b_q * C_pqrs / mu;
                            t1=n_vec(p)*b_vec(q)*cbymu(p,q,r,s);
                            if t1~=0
                                for i=1:2
                                    for k=1:3
                                        for l=1:3
      % t2 = b_i * C_ijkl * G_rk,l
      t2=-1/(16*pi*(1-nu))*b_vec(i)*cbymu(i,j,k,l)* ...
         ((r==k)*3*(a*dx)^2*2*(1-nu)/R_revo^5*x(l)+... % extra term from Cai et al.
         (3-4*nu)*(r==k)/R_revo^3*x(l)+3*x(r)*x(k)*x(l)/R_revo^5- ...
```

```
        ((r==l)*x(k)+(k==l)*x(r))/R_revo^3);
                                          for h=1:2
      int_c_lr(ii,jj,h)=int_c_lr(ii,jj,h)+e(s,j,h)*t2*t1;
                                          end
                                      end
                                  end
                              end
                          end
                      end
                  end
              end
          end
      end
  end
end
end

function [e]=epsi   % Einstein permutation operator (or Levi-Civita symbol)
e=zeros(3,3,3);
for i=1:1:3
    for j=1:1:3
        for k=1:1:3
            if i==1 && j==2 && k==3  % = 1 when i,j,k are in cyclic order
                e(i,j,k)=1;
            elseif i==2 && j==3 && k==1
                e(i,j,k)=1;
            elseif i==3 && j==1 && k==2
                e(i,j,k)=1;
            elseif i==3 && j==2 && k==1  % = -1 when i,j,k in reverse cyclic order
                e(i,j,k)=-1;
            elseif i==1 && j==3 && k==2
                e(i,j,k)=-1;
            elseif i==2 && j==1 && k==3
                e(i,j,k)=-1;
            else
                e(i,j,k)=0;  = 0 when otherwise (i.e. i=j or j=k or k=i)
            end
        end
    end
end
end
```

4.4.2 Applying long-range elastic and core interactions

With the int_c_lr(ii,jj,h) or $g_{\alpha\alpha h}(\vec{R}(ii, jj))$ calculated in the previous subsection, here they are substituted into equation (4.24) to compute the $\tau_\alpha^{\text{int}}(\vec{r})$ (= tau_int(*i*,*j*)) at each meshpoint $\vec{r} = (i,j)$ on the slip plane. The function interaction_lr_core1 below does this. $\rho_\alpha(\vec{r}')$ (= rhoG(*i*,*j*)) is input and ξ_h^α (= *xi*(*i*,*j*,*h*)) is calculated from the input dislocation character (theta(*i*,*j*)) as its cosine (for $h = 1$) and sine (for $h = 2$). The size of the input int_c_lr array is then used to determine the cut-off radius for elastic interaction consideration, and then the neighbours of each meshpoint (*i*,*j*) are located to do the sum in equation (4.24). Periodic boundary conditions are applied so that boundary meshpoints are connected to neighbours on the opposite boundary. The stress-shielding factor f for core interaction is also calculated as per equation (4.25) and the $\tau_\alpha^{\text{int}}(\vec{r})$ is scaled by the factor (1–*f*) as in equation (4.24). The total interaction stress is then returned as the tau_int(*i*,*j*) array.

```
function [tau_int]=interaction_lr_core1(rhoG, theta, int_c_lr, dx, dy, rho_min,
dis_width, core_size, r_screw, n_core, r_n)
% This version is for periodic boundary conditions
% The density Q(i,j) on input is already GND (rhoG) with character N(i,j)
% over spatial points (i,j). theta(i,j) is measured from XX axis, from 0 to
% 2*pi

%Boundary conditions
px=size(rhoG,1); py=size(rhoG,2);
[rho_b]=periodic_2d(rhoG, size(rhoG,1), size(rhoG,2), 1, px, py);
[theta_b]=periodic_2d(theta, size(theta,1), size(theta,2), 1, px, py);

tau_int=zeros(size(rhoG,1),size(rhoG,2));
nx_co = (size(int_c_lr,1)-1)/2;
ny_co = (size(int_c_lr,2)-1)/2;
xi = zeros(size(theta_b,1),size(theta_b,2),2);  % direction cosines of GND
xi(:,:,1) = cos(theta_b(:,:));  % projection of GND onto screw axis
xi(:,:,2) = sin(theta_b(:,:));

d_width=ceil(dis_width/dx);
for i=1:size(rhoG,1)
    for j=1:size(rhoG,2)
        h(i,j) = d_width/2*sqrt((xi(px+i,py+j,1)*r_screw)^2+xi(px+i,py+j,2)^2);
        % screw core narrower than edge core by ratio r_screw
        n(i,j) = n_core*sqrt((xi(px+i,py+j,1)*r_n)^2+xi(px+i,py+j,2)^2);
    end
end

core = core_size*d_width;  % half-width for core shielding in disln normal direction
core_xi = core;            % ditto for the dislocation line direction

% Establishing elastic + core interaction RSS for each field point (i,j)
for i=1:size(rhoG,1)           % i is along horizontal axis, j along vertical axis
    for j=1:size(rhoG,2)
        % Long- and short-range elastic interaction RSS
        for ii=1:nx_co*2+1      % index for int_c_lr, summing over neighbours (ii,jj)
                                % of field point (i,j) within cutoff
            is=px+i+(ii-nx_co-1);  % index for rho, source point x' gradually moves
                                   % from left to right of i as ii increases
            for jj=1:ny_co*2+1
                js=py+j+(jj-ny_co-1);
                if rho_b(is,js)>rho_min
                    % core interaction factor f = (d/h(xi))^n_core
                    d = abs(-(px+i-is)*xi(px+i,py+j,2)+(py+j-js)*xi(px+i,py+j,1));
                    % d is distance along disln normal direction
                    d_xi = abs((px+i-is)*xi(px+i,py+j,1)+(py+j-js)*xi(px+i,py+j,2));
                    % d_xi is distance along disln direction
                    if d<=core && d_xi<=core_xi
                        f = (d/h(i,j))^n(i,j);
                    else
                        f = 0;
                    end
                    tau_int(i,j)=tau_int(i,j)+(int_c_lr(ii,jj,1)*xi(is,js,1)+...
                                 int_c_lr(ii,jj,2)*xi(is,js,2))*rho_b(is,js)*(1-f);
                end              % c.f. eqn. 4.21a
                end
            end
        end
    end
    dV = dx*dy;
    tau_int(:,:)=tau_int(:,:)*dV;
    end
```

The typical call is:

```
V(:,:)=(tau_app + tau_int(:,:))*mobility;
```

4.4.3 Shrinkage of loop under self stress

To demonstrate the effects of the elastic and core interactions, here we present a programme to simulate the shrinkage of an initially circular dislocation loop under its own self stress. The code is as follows and should be self-explanatory:

```
%% DDFD_discretek: for loop expansion/shrinkage with dynamics
% Material properties
b=1;                    % Burgers vector magnitude
b_vec=[1 0 0]*b;        % Burgers vector in the i-j axis system
n_vec=[0 0 1];          % slip plane normal
mu=1;                   % Shear modulus
nu=0.3;                 % Poisson's ratio
mobility = 5;           % dislocation mobility: v = mobility * tau/mu

% Dislocation parameters
Rc_int=50*b;            % cutoff radius of dislocation interactions (100*b)
rho_min=0;              % minimum density to skip long-range interaction (1e-4/b)
a = 1;                  % Cai et al.'s singularity removal length scale
dis_width = 5*b;        % Width of edge dislocation core
core_size = 2;  % Radius for core interaction shielding, in times of (dis_width/dx)
r_screw = 1-nu;         % Ratio of screw width to edge width
n_core = 1.2;           % Strength of core-interaction shielding: n in f = (d/h)^n
r_n = 1-nu;             % Ratio of n_core of screw character to edge character

% Simulation parameters
tau_app = 0*mu;         % Applied RSS on slip system in terms of mu
dt= 1;                  % time step
T=250*dt;               % time to finish
magnitude=100*b;        % size of simulation cell 50b
theta_arc_s = 0;        % starting angle of dislocation arc (from i axis)
theta_arc_f = 2*pi;     % finish angle of dislocation arc (from i axis)
radius_arc= .3*magnitude; % radius of dislocation arc
nxintervals = 101;      % number of x intervals 101
sdev=dis_width*0.4;     % standard deviation for gaussian profile of dislocation core


%% points extraction
x=linspace(0*magnitude,1*magnitude,nxintervals);
y=x;
dx=(x(end)-x(1))/(size(x,2)-1);
dy=(y(end)-y(1))/(size(y,2)-1);

%% Mesh grid (horizontal axis is i, vertical is j)
meshpoints=zeros((size(x,2))^2,2);
k=1;
for i=1:size(x,2)
    for j=1:size(y,2)
        meshpoints(k,1)=x(1,i);
        meshpoints(k,2)=y(1,j);
        k=k+1;
    end
end

%% Establishing the dislocation line
xCenter = magnitude/2;      % circle centre in the meshpoint coordinate system
yCenter = magnitude/2;
xRadius = radius_arc;
yRadius = xRadius;
theta_arc = theta_arc_s : 0.01 : theta_arc_f;
lx = xRadius * cos(theta_arc) + xCenter;
ly = yRadius * sin(theta_arc) + yCenter;
line=[lx;ly];  % a matrix for the discrete points of the line
```

```
%% coarse-graining section
Q=zeros(size(x,2),size(y,2));
QQ=Q; %% QQ is the working density, here initialized to Q
theta = zeros(size(x,2),size(y,2)); % theta(i,j) is dislocation character from XX
[xy,distance,t_a]=distance2curve(line.',meshpoints,'linear');

%% preparing for integration
[int_c_lr]=interaction_coeff(n_vec, b_vec, a, mu, nu, dx, dy, Rc_int);
ds=sqrt(2)*dx;

%% Integration loop
tau_int=zeros(size(QQ,1),size(QQ,2));  % mutual interactive stress between densities
for time=1:round(T/dt)
    tau_int=interaction_lr_core1(QQ, theta, int_c_lr, dx, dy, rho_min, dis_width,
core_size, r_screw, n_core, r_n);
    Vbyds(:,:)=(tau_app + tau_int(:,:))*mobility/ds;  % mobility law
    QQ = difff3(QQ,dt,Vbyds,theta);   % difff3 same as difff2 except input is Vbyds
    %
    % Results can be plotted here, which is skipped for clarity purpose
    %
end
```

Figure 4.9 shows the simulation result of the above programme. As the loop shrinks, it no longer stays circular because the long-range elastic interaction and core interaction are not isotropic, even though elasticity is isotropic. The shape will depend on the input parameters for the material and dislocation, but the fact that the loop shrinks well demonstrates that the Mura interaction is at play. Furthermore, the dislocation width stays rather steady despite the repulsive Mura interaction within the core, indicating that the core-shielding strategy (via the factor $1-f$) works well. The shrinkage is not to the end because we have not built in dipole annihilation in the programme.

4.5 Simulation of Orowan looping

To further illustrate dynamics, Orowan looping is considered here (Kalaei, Xiang and Ngan, 2022). Apart from long-range elastic and core interactions considered above for the shrinking loop, Orowan looping poses two new tasks for simulation: (i) dislocation-particle interaction, and (ii) dipole annihilation (pinch-off), which are discussed in the following two subsections.

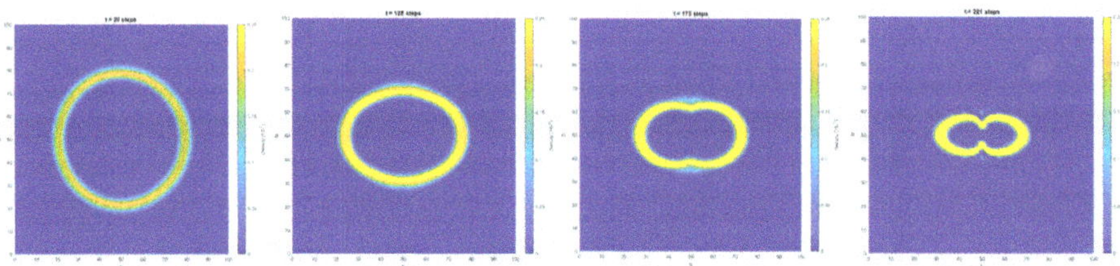

Figure 4.9. Simulation snapshots of an initially circular dislocation loop (leftmost) shrinking under its own stress field.

4.5.1 Dislocation-particle interaction force law

In Orowan looping, a dislocation encounters precipitate obstacles and gets stuck there, while the unobstructed parts of the dislocation bow out. We therefore need a force field to represent the obstruction offered by the precipitates. Real, incoherent precipitates that are non-deformable will bring an approaching dislocation into hard stop at the precipitate interface. However, it goes without saying that hard stop is unfavourable for numerical stability in computer simulations. Therefore, for successful simulation of Orowan looping, we need a particle force field that causes soft stop of the dislocation. The following is one such law:

$$\tau^{\mathrm{obs}}(\vec{r}) = \begin{cases} -\tau_c\left(p + 1 - \dfrac{pR(\vec{r})}{r_p}\right) & \text{for } R < r_p \\ -\tau_c\left(\dfrac{r_p}{R(\vec{r})}\right)^p & \text{for } R \geqslant r_p \end{cases} \tag{4.27}$$

Here, $R(\vec{r})$ is the distance of field point $\vec{r}$ from the particle centre, r_p is a nominal particle radius, and τ_c is a constant stress, which is set close to and slightly higher than the applied resolved shear stress τ^{app} (see equation (4.2)) so that the effective stress $\tau^{\mathrm{obs}} + \tau^{\mathrm{app}} + \tau^{\mathrm{int}}$ would vanish at $R \approx r_p$ for the dislocation to stop at around there. As shown in figure 4.10, τ^{obs} gradually decays outside the particle (with the decay speed controlled by parameter p) and rises linearly on decreasing R inside the particle, with a smooth slope at the boundary $R = r_p$ between the two regimes. This force field may be executed by the following function particle_interaction, which may be called by the following command in the integration loop of the main programme:

```
V=(tau_app + tau_int + tau_particle)*mobility;
```

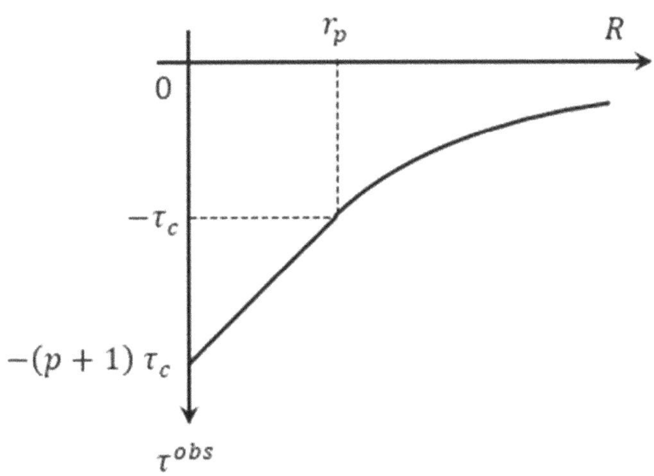

Figure 4.10. Plot of the particle force law for Orowan looping simulation.

```
function [tau_particle]=particle_interaction(XX,YY,xp,yp,Rp,tau_c,mp)
tau_particle=zeros(size(XX,1),size(XX,2));
for i=1:size(XX,1)
    for j=1:size(XX,2)
        r=sqrt((XX(i,j)-xp)^2+(YY(i,j)-yp)^2);
        if r<Rp
            tau_particle(i,j)=-tau_c*(mp+1-mp*r/Rp);
        else
            tau_particle(i,j)=-tau_c*(Rp/r)^mp;
        end
    end
end
end
```

4.5.2 Annihilation

The second issue for Orowan looping is dipole annihilation leading to pinch-off. The function 'annihilation' below executes this using the following procedure:

- For each meshpoint (i,j) with a density Q(i,j) larger than a background level specified by rho_min, the neighbour points (ii,jj) within a cut-off distance r_ann (in terms of pixel number) and with densities also larger than rho_min are identified.
- For each neighbouring point (ii,jj), the dislocation character theta(ii,jj) is compared with the theta(i,j) of the meshpoint under consideration, and if their difference (theta(ii,jj)−theta(i,j)) is close enough to pi within a tolerance margin dtheta_ann, then the neighbour point's density Q(ii,jj) will be incorporated into the average amount of density to be annihilated dQ_ann(i,j) for (i,j).
- After searching through all neighbours of (i,j), the average annihilation amount dQ_ann(i,j) will be deducted from the original density Q(i,j) of the meshpoint.

```
function [Q]=annihilation(Q, theta, r_ann, dtheta_ann)
nb=ceil(r_ann);  % neighbour pixels within which annihilation happens
rho_min = 0;
dQ_ann = zeros(size(Q,1),size(Q,2));  % density to be annihilated
for i=1:size(Q,1)
    for j=1:size(Q,2)
        n_count=0;
        if Q(i,j)<rho_min
            continue
        end
        for ii = i-nb:i+nb       % searching over neigbourhood of (i,j)
            for jj = j-nb:j+nb   % to find dipole densities
                if ii>0 && ii<=size(Q,1) && jj>0 && jj<=size(Q,2) &&
        Q(ii,jj)>=rho_min && abs(abs(theta(ii,jj)-theta(i,j))-pi)<=dtheta_ann
                    n_count=n_count+1;
                    dQ_ann(i,j)=dQ_ann(i,j)+Q(ii,jj);
                end
            end
        end
        if n_count>0, dQ_ann(i,j)=dQ_ann(i,j)/n_count; end  % average dipole density
                                                            % of (i,j) in its neigbourhood
        if dQ_ann(i,j)>Q(i,j), dQ_ann(i,j)=Q(i,j); end
    end
end
Q=Q-dQ_ann;  % doing annihilation
Q(Q<0)=0;    % making sure negative densities are not resulted
end
```

4.5.3 Pinch-off

While the above annihilation function is useful, it is not sufficient to successfully simulate the pinch-off phenomenon of Orowan looping. The reason is that, with the 'annihilation' function, the local densities of opposite signs are annihilated pixel by pixel, leading to local patches remaining that will not be cleaned up. Therefore, pinch-off is handled by a redistribution scheme extended from the re-synchronization method introduced in section 3.4.3.2 in the last chapter.

To recap (see figure 3.11), to handle rotation of dislocation densities (see equation (3.9)), after every few steps of movement by the 'difff' function, the current density distribution $Q(:,:)$ is refitted to a curve and then the $Q(:,:)$ and theta(:,:) are regenerated by the 'distance2curve' and 'discoarse' subroutines (sections 3.4.1 and 3.4.2.2) as Gaussian profiles perpendicular to the fitted curve. The 'redistribute' function introduced in section 3.4.3.2 handles all these. For the present pinch-off operation, the 'redistribute' function may be modified to:

(i) detect whether the newly fitted curve will have dipole segments that are close enough (within a pre-set distance 'rnb') for annihilation to occur, and
(ii) execute the annihilation by breaking the curve at the dipole point and reconnecting the broken curve into two new curves for the pinched-off dislocations, and then calling the 'discourse' function to regenerate the $Q(:,:)$ and theta(:,:) profiles.

These steps are handled by the function 'pinchoff_OL' as illustrated in the flowchart in figure 3.11. The code of this function is given below, and the reader should first refresh the memory of the 'redistribute' function in section 3.4.3.2 in order to understand it. Key features of 'pinchoff_OL' are as follows:

- Unlike in 'redistribute' where the 'trim' factor is pre-set, in 'pinchoff_OL', the trim factor (which trims the Q density for curve fitting) is adjusted automatically instead of set to a fixed value. The adjustment makes sure that the number of points per interval (np) allowed by the trim factor falls within a prescribed range (np_min, np_max), so that the curve-fitting subroutine has less chance of failure. Thus, instead of inputting just one 'trim' factor, the corresponding input is an array fitting(1:3) = [np_min, np_max, trim_initial], where trim_initial is the initial value of the trim factor.
- The function decides on a flag 'pinch_off', which is a logical variable. In the main programme before the integration loop, this variable is set to false. Then every time 'pinchoff_OL' is called, the function decides whether pinch-off can happen—if it does happen, then 'pinch_off' is reset to true, otherwise it is left as false. The 'pinch_off' variable instructs the subsequent flow in the main programme (see figure 3.11) according to whether pinch-off has occurred or not.

- If pinch-off is detected from the curve fitted to the input *Q*(:,:), then, as mentioned above, the fitted line is broken at the dipole point (see figure 4.11), and the fitting points (*xQ*(:), *yQ*(:)) are reconnected into two new curves: (1) for the initial dislocation, (2) for the pinched-off loop around the precipitate. These two curves are then respectively sent to the 'distance2curve' and 'discourse' functions for regenerating their respective density and dislocation-character profiles: Q1(:,:) and theta1(:,:) for dislocation 1; Q2 (:,:) and theta2(:,:) for dislocation 2, which are returned to the main programme. The average number of points on the fitted curves per pixel, np1 and np2, are also returned to the main programme for monitoring of the curve-fitting quality.

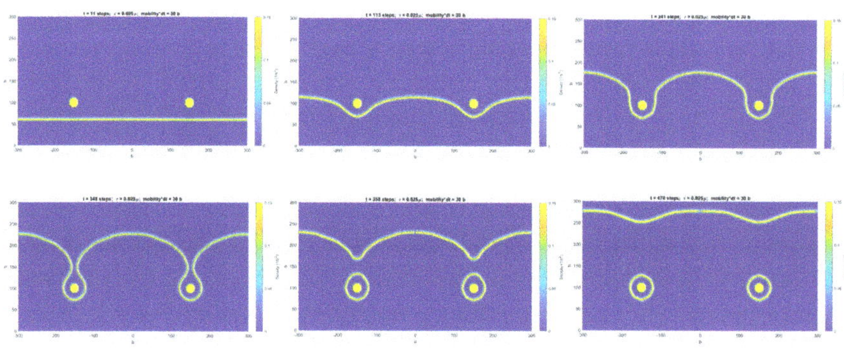

(a)

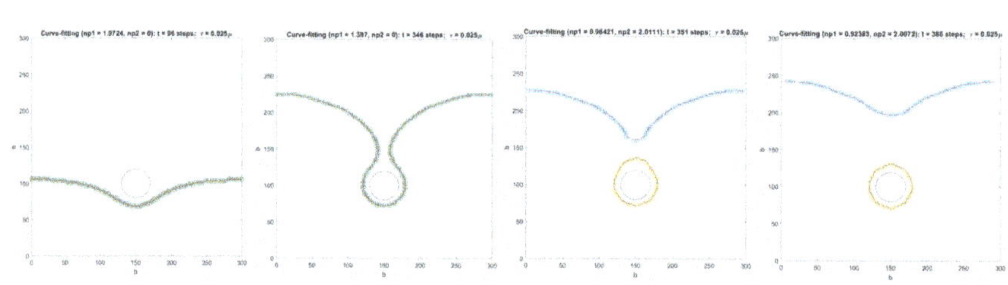

(b)

Figure 4.11. Simulation snapshots of Orowan looping. (a) Plots of dislocation density Q in the simulation cell and its mirror image about the vertical axis. To show the dislocation and particles more clearly, the particles are drawn with a radius smaller than the nominal value (r_p or *Rp*, section 4.5.1) by 1.5 times the dislocation width (dis_width). (b) Plots of the fitted curve(s) and points on the trimmed *Q*(:,:) profile used for the curve fitting. The middle two figures are snapshots just before and after pinch-off. The precipitate in (b) is drawn with its nominal radius r_p or *Rp*.

```
function
[pinched_off,Q1,Q2,theta1,theta2,h_a1,h_a2,lx1,lx2,ly1,ly2,xQ1,xQ2,yQ1,yQ2,np1,np2]=p
inchoff_OL(pinched_off,Q,theta,meshpoints,XX,YY,magnitude,h_a,x,sdev,p,dx,skip,rnb,fi
tting)
%% To detect and execute pinch off (annihilation) for Orowan looping
%  Pinch-off of Orowan looping is by auto-detection when two points along the fitted
dislocation curve is closer
%  than rnb. Then the original dislocation is split into two which are stored
separately as densities Q1(i,j) and
%  Q2(i,j), and orientation theta1(i,j) and theta2(i,j), over the common slip-plane
domain

%% To first trim the input density Q to obtain a more well-defined (narrow)
dislocation profile for curve fitting
Q(Q<0)=0;
QSum0=sum(sum(Q));
Qmax=max(Q,[],'all','linear');
Qmin=fitting(3)*Qmax;   % fitting(3) is initial trim factor; later adjusted
automatically

%% Fit curve to the trimmed density field Q
np=0;         % points per dx
while np<fitting(1) || np>fitting(2)   % fitting(1) and (2) are limits of points per
dx
    [xQ,yQ,ha,n]=search_points(Q,XX,YY,h_a,Qmin);
    if np==0 && n<=fitting(1)*magnitude
        Qmin=Qmin*0.99;
        continue
    elseif np==0 && n>=fitting(2)*magnitude
        Qmin=Qmin*1.01;
        continue
    end
    [~,ha_sort]=sort(ha);   % ha_sort is the order of ha, the fractional arc length
                            % from the start of dislocation
    xQ=xQ(ha_sort); % xQ and yQ are sorted according to their fractional arc length
    yQ=yQ(ha_sort);
    ln=1:1:n;
    lx=csaps(ln,xQ,p,ln);
    ly=csaps(ln,yQ,p,ln);
    arc_length=integrate(lx,ly,n)/dx;   % arc length in terms of dx
    np=n/arc_length;                    % points per dx
    if np<fitting(1)
        Qmin=Qmin*0.99;
    elseif np>fitting(2)
        Qmin=Qmin*1.01;
    end
end

%% Detect whether fitted curve has self-touching points within rnb
for i=1:ceil(size(lx,2)/2)  % finding pinch-off point-pair that are half the
                            % dislocation length away
    for j=i+ceil(size(lx,2)/10):size(lx,2)
        r=sqrt((lx(j)-lx(i))^2+(ly(j)-ly(i))^2);
        if r<=rnb
            pinched_off=true;
```

```
            n1=i; n2=j;
            break;
        end
    end
end

%% If pinch-off hasn't happened, return to main programme to move dislocation further
if ~pinched_off    % these values are returned as dummies
    lx1=lx; ly1=ly; xQ1=xQ; yQ1=yQ;
    lx2=lx; ly2=ly; xQ2=xQ; yQ2=yQ;
    h_a1=h_a; h_a2=h_a; Q1=Q; Q2=Q; theta1=theta; theta2=theta;
    np1=np; np2=0;
    return;
end


%% Execute pinch-off here
%% Forming dislocation 1 (the original dislocation)
length1=n1+size(lx,2)-n2-2*skip+1;    % number of points on dislocation 1
xQ1=zeros(1,length1); yQ1=zeros(1,length1);
n=0;
for k=1:n1-skip   % first part of dislocation 1
    n=n+1;
    xQ1(n)=xQ(k); yQ1(n)=yQ(k);
end
for k=n2+skip:size(lx,2)   % second part of dislocation 1
    n=n+1;
    xQ1(n)=xQ(k); yQ1(n)=yQ(k);
end


% Apply periodic boundary conditions along x for curve fitting within the cell range
px=length1;
xQ1_c=cumulative_1d(xQ1, size(xQ1,2), -magnitude, px);
yQ1_c=periodic_1d(yQ1, size(yQ1,2), px);
ln1_c=1:1:length1+2*px;
lx1_c=csaps(ln1_c,xQ1_c,p,ln1_c);
ly1_c=csaps(ln1_c,yQ1_c,p,ln1_c);
lx1=zeros(1,length1);
ly1=zeros(1,length1);
lx1(1:length1)=lx1_c(px+1:px+length1);
ly1(1:length1)=ly1_c(px+1:px+length1);

arc_length=integrate(lx1,ly1,length1)/dx;   % arc length in terms of dx
np1=length1/arc_length;


line=[lx1;ly1];
% coarse-graining section
theta1 = zeros(size(Q,1),size(Q,2));
Q1 = zeros(size(Q,1),size(Q,2));
[~,distance,t_a]=distance2curve(line.',meshpoints,'linear');
[Q1,theta1,h_a1]= discoarse2(distance,t_a,sdev,x,line,Q1,theta1);

%% Forming dislocation 2 (the pinched-off loop around the precipitate)
length2=n2-n1-2*skip+2; % number of points on dislocation 2
xQ2=zeros(1,length2); yQ2=zeros(1,length2);
n=0;
```

```
ns=ceil((n1+n2)/2);      % take mid-point of dislocation 2 as new start/finish points,
for k=ns:n2-skip         % instead of having the pinched-off points as start/finish
                         % points, to achieve smoothness at the cusp
    n=n+1;
    xQ2(n)=xQ(k); yQ2(n)=yQ(k);
end
for k=n1+skip:ns
    n=n+1;
    xQ2(n)=xQ(k); yQ2(n)=yQ(k);
end

ln=1:1:length2; % re-curve-fitting coordinates of dislocation 2 to improve smoothness
lx2=csaps(ln,xQ2,p,ln);
ly2=csaps(ln,yQ2,p,ln);

arc_length=integrate(lx2,ly2,length2)/dx;   % arc length in terms of dx
np2=length2/arc_length;


line=[lx2;ly2];
% coarse-graining section
theta2 = zeros(size(Q,1),size(Q,2));
Q2 = zeros(size(Q,1),size(Q,2));
[~,distance,t_a]=distance2curve(line.',meshpoints,'linear');
[Q2,theta2,h_a2]= discoarse2(distance,t_a,sdev,x,line,Q2,theta2);

%% Normalizing the dislocaton densities to preserve the same sum as the input
QSum=sum(sum(Q1))+sum(sum(Q2));
Q1=Q1*QSum0/QSum;
Q2=Q2*QSum0/QSum;
end


function [arc_length]=integrate(lx,ly,n)
arc_length=0;
for i=1:n-1
    arc_length=arc_length+sqrt((ly(i+1)-ly(i))^2+(lx(i+1)-lx(i))^2);
end
end


function [xQ,yQ,ha,n]=search_points(Q,XX,YY,h_a,Qmin)
% To search for points that meet Q>=Qmin for curve fitting
n=0;
for i = size(Q,1):-1:1
    for j = 1:size(Q,2)
        if Q(i,j)>=Qmin
            n = n+1;
        end
    end
end
xQ=zeros(1,n); yQ=zeros(1,n); ha=zeros(1,n);
k=0;
for i = size(Q,1):-1:1
    for j = 1:size(Q,2)
        if Q(i,j)>=Qmin
            k = k+1;
            xQ(k)=XX(i,j);
```

```
                yQ(k)=YY(i,j);
                ha(k)=h_a(j,i);
            end
        end
    end
end

function [mat_p]=cumulative_1d(mat, m, cell, px)
% To apply cumulative boundary condition
if m>=px
    mat_p=zeros(1,m+px*2);
    mat_p(1+px:m+px)=mat;   % central cell
    mat_p(1:px)=mat(m-px+1:m)-cell;  % left side
    mat_p(m+px+1:m+2*px)=mat(1:px)+cell; % right side
end
end

function [mat_p]=periodic_1d(mat, m, px)
% To apply periodic boundary condition
if m>=px
    mat_p=zeros(1,m+px*2);
    mat_p(1+px:m+px)=mat;   % central cell
    mat_p(1:px)=mat(m-px+1:m);  % left side
    mat_p(m+px+1:m+2*px)=mat(1:px); % right side
end
```

4.5.4 Simulation results

Finally, we are ready to see how the dislocation interactions, particle force field, annihilation, and pinch-off subroutines work together to simulate Orowan looping. The following shows extracts of the main-programme code; commands identical to the DDFD_discretek programme in section 4.4.3 are omitted here to improve clarity. The reader should also refer back to figure 3.11 to understand the flow. After pinch-off has occurred, the programme moves and rotates the two dislocations separately with different control parameters so as to provide more versatile control of the stability. The total density and theta fields are given by superposition and weighted averaging:

```
Q(:,:) = Q1(:,:) + Q2(:,:)

theta(:,:) = (theta 1(:,:)*Q1(:,:) + theta 2(:,:)*Q2
(:,:))/Q(:,:),
```

for the purposes of interaction calculation by the ‘interaction_lr_core1’ function, as well as plotting.

```
%% DDFD_Orowan_Looping_discretek: for Orowan looping
:
% Simulation parameters
tau_app = 0.025*mu;      % Applied RSS on slip system in terms of mu
equi_time = 10;          % initial equilibration time before stress application

% Stability control parameters
p = 0.1;                 % (in redistribute and pinchoff) smoothness of curve fitting
                         % in MATLAB function csaps (lower p => higher smoothness)
fitting1 =  [1 2 0.9];   % (in redistribute and pinchoff) [min, max no of points per
                         % dx, initial trim] for curve-fitting (set around 2) before
                         % pinch off
fitting2_d1=[0.5 1 0.9];% ditto for disln 1 after pinch off. Set point no. to low to
                         % avoid over-fitting, and vice versa
fitting2_d2=[2 4 0.9];   % ditto for disln 2 after pinch off. Set initial trim to high
                         % (e.g. 0.9) to be close to lower limit for point no., and
                         % vice versa
skip = 5;                % (in pinchoff) number of points skipped on either broken
                         % segments to make annihilation happen during pinch off
                         % (higher means smaller pinched-off cusps)
rnb = dis_width*3;       % (in pinchoff) neighbour distance for annihilation to occur
r_ann=dis_width/dx*3;    % (in annihilation) neighbour distance for annihilation
dtheta_ann = 0.2;        % (in annihilation) dipole misorientation angle (in radians)
                         % for annihilation to occur
redis_freq1 = 2;         % number of advancement steps before each redistribution,
                         % before pinch-off
redis_freq2_d1 = 2;      % number of advancement steps before each redistribution for
                         % dislocation 1, after pinch-off
redis_freq2_d2 = 2;      % number of advancement steps before each redistribution for
                         % dislocation 2, after pinch-off

%% Setting up the particle
Rp=20*b;                 % nominal radius of particle force field
tau_c=tau_app*2.5;       % characteristic stress in particle force field
mp=2;                    % exponent in decaying part of particle force law outside Rp:
                         % tau=tau_c*(r/Rp)^(-mp)
L = magnitude/2;         % particle half-spacing
xp=L;                    % particle position in the meshpoint coordinate system
yp=yCenter-L/3;

% Extra lines in main programme before integration loop
XX=zeros(size(x,2),size(x,2));
YY=XX;
for i=1:size(x,2)
    for j=1:size(y,2)
        XX(i,j)= x(1,i);
        YY(i,j)= y(1,j);
    end
end
tau_particle=particle_interaction(XX,YY,xp,yp,Rp,tau_c,mp);
pinched_off=false;
Qmax0=max(Q,[],'all','linear');

%% Integration loop
for time=1:round(T/dt)+equi_time
```

```
    if time<= equi_time       % equilibration without applied stress
        tau_int=interaction_lr_core1(Q, theta, int_c_lr, dx, dy, rho_min,
                ...dis_width, core_size, r_screw, n_core, r_n);
        V=mobility*tau_int;
        [~,rot]=rotation1(theta,Q,V,dt,dx,dy,dis_width,rho_min);
        theta=theta+rot;
        Q = difff3(Q,dt,V/ds,theta);  % difff3 uses Vbyds as input
        % Cleaning and renormalization
        Q(Q<rho_min)=0;
        Qmax=max(Q,[],'all','linear');
        Q=Q*Qmax0/Qmax;
    elseif ~pinched_off      % move dislocation if pinch-off hasn't occurred
        tau_int=interaction_lr_core1(Q, theta, int_c_lr, dx, dy, rho_min,
                 ... dis_width, core_size, r_screw, n_core, r_n);
        V=(tau_app + tau_int + tau_particle)*mobility;
        [~,rot]=rotation1(theta,Q,V,dt,dx,dy,dis_width,rho_min);
        theta=theta+rot;
        Q = difff3(Q,dt,V/ds,theta);  % difff3 uses Vbyds as input
        % Cleaning and renormalization
        Q(Q<rho_min)=0;
        Qmax=max(Q,[],'all','linear');
        Q=Q*Qmax0/Qmax;
        % Redistribute density and re-calculate theta by fitting line to density
        if mod(time,redis_freq1)==0
            [Q,theta,h_a,lxx,lyy,xQ,yQ,~]=redistribute_periodic(Q,meshpoints,...
                                           XX,YY,magnitude,h_a,x,sdev,p,dx,fitting1);
        end
        % detect and execute pinch-off if this has not yet happened
[pinched_off,Q1,Q2,theta1,theta2,h_a1,h_a2,lx1,lx2,ly1,ly2,xQ1,xQ2,yQ1,yQ2,np1,np2]=p
inchoff_OL(pinched_off,Q,theta,meshpoints,XX,YY,magnitude,h_a,x,sdev,p,dx,skip,rnb,fi
tting1);
        if pinched_off, time_pinched_off=time; end
    elseif pinched_off      % execute this section after pinch-off; there are now 2
dislocations to move
        % elastic interactions are calculated for the overall Q=Q1+Q2
        Q=Q1+Q2;
        theta=(theta1.*Q1+theta2.*Q2)./Q;   % theta field is weighted average of
theta1 and theta2
        theta(isnan(theta))=0;              % removing singular elements
        tau_int=interaction_lr_core1(Q, theta, int_c_lr, dx, dy, rho_min, ...
                 dis_width, core_size, r_screw, n_core, r_n);
        V=(tau_app + tau_int + tau_particle)*mobility;

        % moving dislocation 1 (the original dislocation after by-passing particle)
        [~,rot]=rotation1(theta1,Q1,V,dt,dx,dy,dis_width,rho_min);
        theta1=theta1+rot;
        [Q1]=annihilation(Q1,theta1,r_ann,dtheta_ann);
        Q1 = difff3(Q1,dt,V/ds,theta1);  % difff3 uses Vbyds as input
        % Cleaning and renormalization
        Q1(Q1<rho_min)=0;
        Qmax=max(Q1,[],'all','linear');
        Q1=Q1*Qmax0/Qmax;

        % moving dislocaton 2 (the pinched-off loop)
        [~,rot]=rotation1(theta2,Q2,V,dt,dx,dy,dis_width,rho_min);
```

```
        theta2=theta2+rot;
        [Q2]=annihilation(Q2,theta2,r_ann,dtheta_ann);
        Q2 = difff3(Q2,dt,V/ds,theta2);  % difff3 uses Vbyds as input
        % Cleaning and renormalization
        Q2(Q2<rho_min)=0;
        Qmax=max(Q2,[],'all','linear');
        Q2=Q2*Qmax0/Qmax;

        % Redistribute density and re-calculate theta by fitting line to density
        if mod(time-time_pinched_off,redis_freq2_d1)==1
[Q1,theta1,h_a1,lx1,ly1,xQ1,yQ1,np1]=redistribute_periodic(Q1,meshpoints,XX,YY,magnit
ude,h_a1,x,sdev,p,dx,fitting2_d1); % Similar to redistribute except for periodic b.c.
            [Q1]=annihilation(Q1,theta1,r_ann,dtheta_ann);
        end
        if mod(time-time_pinched_off,redis_freq2_d2)==1
             [Q2,theta2,h_a2,lx2,ly2,xQ2,yQ2,np2]=redistributeLoop(Q2,meshpoints,...
                                        XX,YY,h_a2,x,sdev,p,dx,fitting2_d2);
             % Similar to redistribute except to join cut points to make loop
            [Q2]=annihilation(Q2,theta2,r_ann,dtheta_ann);
        end

        Q=Q1+Q2;
        theta=theta1+theta2;
    end
    tau_int=interaction_lr_core1(Q, theta, int_c_lr, dx, dy, rho_min,...
            dis_width, core_size, r_screw, n_core, r_n);
    %
    % Results can be output here, but this is skipped for clarity
    %
end
```

Figure 4.11 shows simulation snapshots from the above code. The simulation cell contains only one precipitate at the centre with periodic boundary conditions applied but in figure 4.11(a), for better visual effects, the cell and its mirror image about the vertical axis are shown to show the bow out between two precipitates. The curve-fitting results are shown in figure 4.11(b). From the middle two snapshots just before and after the pinch-off, we can see how the original curve is broken and reconnected into two curves.

4.6 Simulation of Frank–Read source

As one further application, the simulation strategies above for Orowan looping equally apply to Frank–Read source, which also involves annihilation and pinch-off. There is no more interacting particle in a Frank–Read source, but then we have to develop a method for effectively pinning the source at its two ends. Again, any hard stop or sharp discontinuity will be unfavourable for numerical stability, so applying a point force at the pin to nullify the effective stress τ^{eff} there would not work. Instead, we should introduce smooth gradients in the velocity profile along the Frank–Read source near the pinning points. This may be done by scaling the velocity $v(\vec{r})$ calculated from the effective stress via equation (4.2) in the following way:

$$v(\vec{r}) \to v(\vec{r}) \times [4t(\vec{r})(1 - t(\vec{r}))]^p \tag{4.28}$$

Here, $t(\vec{r})$ is the fractional arc length, i.e., the t_a returned by the 'distance2curve' subroutine (section 3.4.1), that varies from 0 at one pin to 1 at the other pin of the Frank–Read source. The scaling factor $[4t(\vec{r})(1 - t(\vec{r}))]^p$ therefore is 1 at the mid-point of the dislocation line where $t = 0.5$, and drops to zero at the two ends. The gradient is controllable by the parameter p. The following simple function executes equation (4.28) for a p value of 0.2.

```
function [V]=velocityFR(h_a, tau_app, tau_int, mobility)
V(:,:)=(tau_app + tau_int(:,:))*mobility;
h_a(h_a>1)=1;
h_a(h_a<0)=0;
% set the velocities at the two ends of the dislocation (i.e. the pins) to zero
for i = 1:size(h_a,2)
    for j = 1:size(h_a,1)
        r=h_a(j,i);
        V(i,j) = V(i,j)*(4*r*(1-r))^0.2;
    end
end
end
```

While the above function can pin the Frank–Read source at its two ends, when used together with the re-synchronization protocol in section 3.4.3.2, the two ends of the Frank–Read source may drift slowly, due to the slight loss of density there after each cycle of density regeneration. To overcome this problem, the density at the two pins can be replenished after each call of the regeneration. The following function 'addpin' executes this.

```
function [Q]=addpin(Q,XX,YY,xPinL,yPinL,xPinR,yPinR,pin_radius)
% Due to the trimming by redistribute, the densities at the pinning points may be
% lost and these are compensated by adding densities back at the pins.

Qmax=max(Q,[],'all','linear');
for i=1:size(Q,1)
    for j=1:size(Q,2)
        rL=sqrt((XX(i,j)-xPinL)^2+(YY(i,j)-yPinL)^2);
        rR=sqrt((XX(i,j)-xPinR)^2+(YY(i,j)-yPinR)^2);
        if rL<=pin_radius || rR<=pin_radius
            Q(i,j)=Qmax;
        end
    end
end
end
```

The main programme for Frank–Read source simulation is quite similar to that for Orowan looping discussed in the last section, except for the addition of the velocity scaling by 'velocityFR' and pin-density replenishment by 'addpin', and so to save space, it is not given here. Figure 4.12 shows simulation snapshots which demonstrate the smooth evolution of the Frank–Read source, and the pinch-off process.

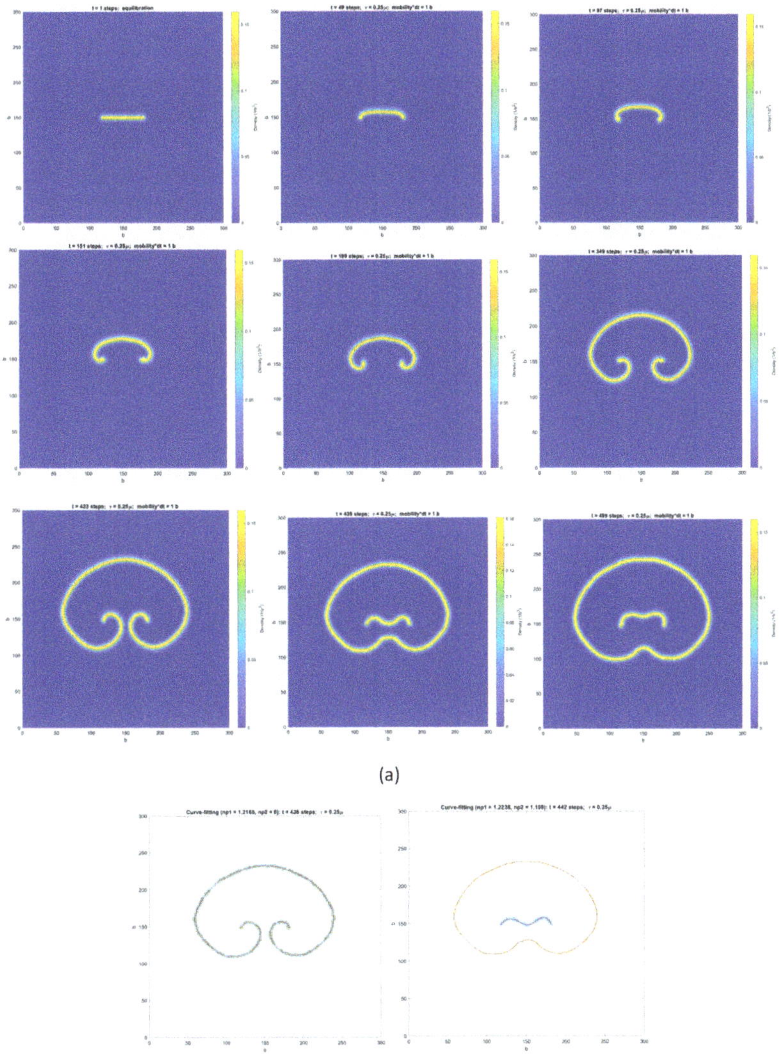

Figure 4.12. Simulation snapshots of operation of a Frank–Read source 60*b* long under an applied resolved shear stress of 0.25 μ. (a) Plots of dislocation density Q. (b) Plots of the fitted curve(s) and points on the trimmed $Q(:,:)$ profile used for the curve fitting, just before and after pinch-off.

References

Cai W, Arsenlis A, Weinberger C R and Bulatov V V 2007 A non-singular continuum theory of dislocations *J. Mech. Phys. Solids* **54** 561–87

Hirth J P and Lothe J 1992 *Theory of Dislocations* 2nd edn (Malabar, FL: Krieger Publishing)

Kalaei A, Xiang Y and Ngan A H W 2022 An efficient and minimalist scheme for continuous dislocation dynamics *Int. J. Plast.* **158** 103433

Mura T 1987 *Micromechanics of Defects in Solids* 2nd edn (Dordrecht: Martinus Nijhoff Publishers, Kluwer Academic)

Yonenaga I 2013 Dislocation dynamics in SiGe alloys *J. Phys: Conf. Ser.* **471** 012002

Zhang Y and Ngan A H W 2018 Dislocation–density dynamics for modeling the cores and Peierls stress of curved dislocations *Int. J. Plast.* **104** 1–22

Zhao S, Osetsky Y N and Zhang Y 2017 Atomic-scale dynamics of edge dislocations in Ni and concentrated solid solution NiFe alloys *J. Alloys Compd.* **701** 1003–8

Chapter 5

All-dislocation density (ADD) for extensive microstructures—kinematics and dynamics closure

5.1 Extensive microstructures and ADD revisited

In chapters 3 and 4, we have examined the dynamics of line densities representing *intensive* microstructures. To recap our discussion in section 2.4, an intensive microstructure is a single dislocation or a bundle of parallel dislocations represented by a vector-density function $\vec{\rho}(\vec{r})$ for which the exact kinematics law is given in various equivalent forms, namely, equations (3.4), (3.9), (3.11) and (3.13). For $\vec{\rho}$ to be a meaningful representation, the dislocation contents must align in the same direction of the vector $\vec{\rho}$, hence either inside a single dislocation or a bundle of parallel dislocations.

In addition to intensive microstructure, we have also said in section 2.4 that an important length scale of CDD is the meso-scale at which the microstructures become *extensive*. A typical resolution pixel in this length scale over which the density is defined would be a good fraction of a micron, with easily $\sim 10^2$ dislocations inside. This amount of dislocations will unlikely have the same line direction and be representable by a single $\vec{\rho}$; hence our treatments in chapters 3 and 4 would become inapplicable. In section 2.3.1, we defined the GND density $\vec{\rho}_G$ as well as the SSD density ρ_S, being, respectively, the net sum and the dipole content of the dislocation-density vectors inside a volume pixel ΔV. At this juncture, a pertinent question to ask is: does the kinematics law (3.4) apply to the GND density $\vec{\rho}_G$? Acharya and Roy (2006) showed that the answer is no, and their argument is as follows. For dislocations of the same parallel bundle i, they obey the kinematics law (3.4) as $\dot{\vec{\rho}}_i = -\vec{\nabla} \times (\vec{\rho}_i \times \vec{v}_i)$. However, from the definition of $\vec{\rho}_G$ in equation (2.13),

$$\dot{\vec{\rho}}_G = \sum_{i\in\Delta V} \dot{\vec{\rho}}_i = -\sum_{i\in\Delta V} [\vec{\nabla} \times (\vec{\rho}_i \times \vec{v}_i)] = -\vec{\nabla} \times \sum_{i\in\Delta V} (\vec{\rho}_i \times \vec{v}_i) \neq -\vec{\nabla} \times \left[\left(\sum_{i\in\Delta V} \vec{\rho}_i\right) \times \vec{v}_G\right],$$

i.e.,

$$\dot{\vec{\rho}}_G \neq -\vec{\nabla} \times (\vec{\rho}_G \times \vec{v}_G) \tag{5.1}$$

where $\vec{v}_G$ is an average velocity of the $\vec{\rho}_i$'s, e.g., $\vec{v}_G = (\sum_{i\in\Delta V} |\vec{\rho}_i|\vec{v}_i)/\rho_{total}$. The Maxwell law therefore does not apply to $\vec{\rho}_G$, and in fact, no exact kinematics law for $\vec{\rho}_G$ is known to exist as of today. Acharya and Roy proposed an error term $\vec{L}^p = \sum_{i\in\Delta V} (\vec{\rho}_i \times \vec{v}_i) - (\vec{\rho}_G \times \vec{v}_G)$, which if added to the GND flux $(\vec{\rho}_G \times \vec{v}_G)$ would make the equal sign in equation (5.1) hold:

$$\dot{\vec{\rho}}_G = -\vec{\nabla} \times (\vec{\rho}_G \times \vec{v}_G + \vec{L}^p) \tag{5.2}$$

$\vec{L}^p$ has the meaning of the SSD flux, but still, there is no exact governing equation for it.

Since no exact kinematics law exits for $\vec{\rho}_G$ or $\vec{L}^p$, any attempt to build a CDD model for extensive microstructures based on GNDs and SSDs would not be entirely satisfactory. However, we still have the exact kinematics law (3.4) for the $\vec{\rho}_i$ for each of the parallel dislocation bundles i within the coarse-grained pixel ΔV, and this suggests that the coarse-grained ADD, defined in equation (2.14), may be a way out. Therefore, in the following section, we will derive an exact and yet simple kinematics law for the ADD $\varrho(\vec{r}, \theta)$ introduced in section 2.3.2. This law was first derived by one of the present authors (Ngan 2017).

5.2 ADD and kinematics law of individual dislocations

As discussed in section 2.3.2, the ADD is defined on a coarse-graining level whereby each volume pixel would consist of many individual dislocations while their effects can be represented by a smeared-out density $\varrho(\vec{r}, \theta)$ over both the real spatial domain and the dislocation character domain. It then follows that, for a kinematics law of the ADD to make sense and be rigorous, the contribution of each individual dislocation within the ADD must be clearly defined and delineated. Therefore, in this section, we aim to develop the coarse-graining scheme of individual dislocations, and the kinematics law for their coarse-grained densities. After this step is done, we can then sum up the effects of individual dislocations to obtain the kinematics law for the entire ADD.

Figure 5.1 shows a single dislocation i moving on a slip system. The ADD due to this dislocation alone may be defined and explained as follows:

$$\varrho_i(\vec{r}, \theta) = \frac{\delta_r(r_i')}{b}\frac{\delta_\theta(\theta_i')}{2\pi} \tag{5.3}$$

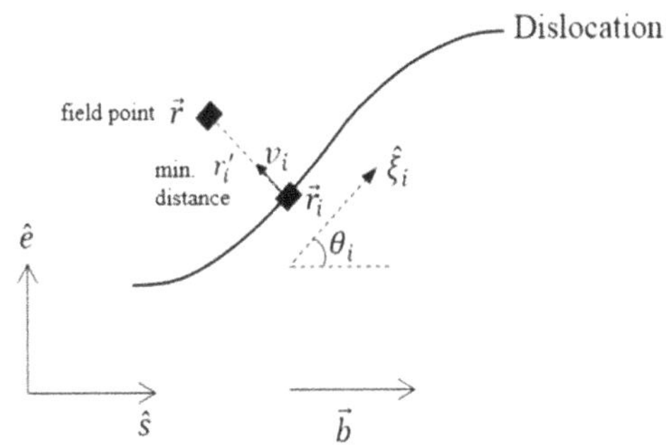

Figure 5.1. An individual dislocation i moving on a slip system defined by Burgers vector $\vec{b}$ and slip-plane normal $\hat{n}$. $\vec{r}$ is a nearby field point, $\vec{r}_i$ is the point on the dislocation that is closest to $\vec{r}$, and $r_i' = |\vec{r} - \vec{r}_i|$ is their separation. v_i is the local dislocation speed, $\hat{\xi}_i$ the local dislocation line direction, and θ_i the local dislocation character angle, at $\vec{r}_i$.

- $\delta_r(r_i')/b$ is the core shape of the dislocation, with $\delta_r(r_i')$ being a delta-like (or narrow Gaussian-like) function defined on coordinate r_i', which is the shortest distance of the dislocation from the field point $\vec{r}$, as shown in figure 5.1. Therefore, when dislocation i is situated far from the field point $\vec{r}$, no dislocation density will be felt at $\vec{r}$ since $\delta_r(r_i')$ will approach zero. When the dislocation is close enough to $\vec{r}$, a perpendicular can be drawn from $\vec{r}$ to the dislocation line to locate the point of closest proximity $\vec{r}_i$ on the latter as shown in figure 5.1, and then the distance $r_i' = |\vec{r} - \vec{r}_i|$ can be well defined. When r_i' is small enough, $\delta_r(r_i')$ is non-zero and then a finite dislocation density may be felt at the field point $\vec{r}$. $\delta_r(r_i')$ satisfies the normalization requirement $\int_{-\infty}^{\infty} \delta_r(r_i')dr_i' = 1$, and the b here is a length scale (close to the Burgers vector magnitude) that represents the thickness of the slip plane, so that the full density of dislocation i equals $\int_{-\infty}^{\infty} \delta_r(r_i')dr_i'/b = 1/b$, i.e., there is one dislocation over the entire slip plane.
- $\delta_\theta(\theta_i')/2\pi$ represents the θ-dependence of the $\varrho_i(\vec{r}, \theta)$ function, where the argument is defined as $\theta_i' = \theta - \theta_i$, with θ_i being the character of the dislocation at the point $\vec{r}_i$ of closest proximity to the field point $\vec{r}$, as shown in figure 5.1. Here, $\delta_\theta(\theta_i')$ is another delta-like function and its physical meaning is this: when the field point $\vec{r}$ is close to the dislocation so that it registers a non-zero $\delta_r(r_i')$, then the ADD $\varrho_i(\vec{r}, \theta)$ should be non-zero only if the θ variable is close to the character angle θ_i of the dislocation. Therefore, $\delta_\theta(\theta_i')/2\pi$ is the spread of the dislocation density in the θ-space, similar to $\delta_r(r_i')/b$ being the spread of the density in the $\vec{r}$-space. $\delta_\theta(\theta_i')$ satisfies the normalization condition $\int_0^{2\pi} \delta_\theta(\theta_i')d\theta_i' = 2\pi$ so that the full content of the dislocation is within the domain $[0, \ 2\pi]$ in which θ is defined.

Next, we proceed to develop the kinematics law of the $\varrho_i(\vec{r}, \theta)$ for dislocation i. To do this, we just perform differentiation of equation (5.3) with respect to time t:

$$\dot{\varrho}_i(\vec{r}, \theta) = \left.\frac{\partial \varrho_i}{\partial t}\right|_{(\vec{r}, \theta)} = \frac{\delta_\theta(\theta_i')}{2\pi b}\delta_r'\left(r_i'\right)\left.\frac{\partial r_i'}{\partial t}\right|_{(\vec{r}, \theta)} + \frac{\delta_r(r_i')}{2\pi b}\delta_\theta'\left(\theta_i'\right)\left.\frac{\partial \theta_i'}{\partial t}\right|_{(\vec{r}, \theta)} \tag{5.4}$$

where $\delta_r'(r_i') = d\delta_r(r_i')/dr_i'$ and $\delta_\theta'(\theta_i') = d\delta_\theta(\theta_i')/d\theta_i'$. In equation (5.4), $\partial/\partial t|_{(\vec{r}, \theta)}$ means that the partial differentiation with respect to t is to be carried out at fixed $\vec{r}$ and θ. Therefore, as dislocation i moves and tilts on the slip plane,

$$\left.\frac{\partial r_i'}{\partial t}\right|_{(\vec{r}, \theta)} = -v_i$$

and

$$\frac{\partial \theta_i'}{\partial t} = \left.\frac{\partial(\theta - \theta_i)}{\partial t}\right|_{(\vec{r}, \theta)} = -\frac{\partial \theta_i}{\partial t} = -\omega_i$$

where v_i and ω_i are, respectively, the linear and rotational speed of the dislocation at the point of closest proximity $\vec{r}_i$ (see figure 5.1). Equation (5.4) then becomes:

$$\dot{\varrho}_i(\vec{r}, \theta) = \underbrace{-\frac{\delta_\theta\left(\theta_i'\right)}{2\pi b}v_i\,\delta_r'\left(r_i'\right)}_{\text{approach}} \underbrace{-\frac{\delta_r\left(r_i'\right)}{2\pi b}\omega_i\,\delta_\theta'\left(\theta_i'\right)}_{\text{tilting}} \tag{5.5}$$

which is the kinematics law for the ADD of dislocation i. The law consists of two terms, one due to the linear approach of the dislocation to the field point $\vec{r}$, and the other due to the rotation of the dislocation as it is in the proximity of the field point $\vec{r}$. Both the linear approach and the rotation of the dislocation contribute to the change in the ADD, since the latter is defined in both the $\vec{r}$-domain as well as the θ-domain. The reader will recall equation (3.9) in chapter 3, which also comprises a movement and a tilting term, albeit for the vector-density $\vec{\rho}$ for a dislocation bundle. The inter-relation between the kinematics laws for the ADD ϱ and the vector-density $\vec{\rho}$ will be discussed in a later section.

5.3 Kinematics law of ADD for extensive microstructures comprising many dislocations

For the extensive coarse-graining level where the volume pixel consists of many dislocations of different line directions (but still in the same slip system of $\{\vec{b}, \hat{n}\}$), the kinematics law of the corresponding ADD can be obtained easily by summing up the contributions from the constituent dislocations i within the representative pixel volume ΔV in equation (5.3). The following averaged, coarse-grained quantities can be defined:

$$\text{Coarse} - \text{grained ADD:}\;\; \varrho(\vec{r}, \theta) := \sum_{i \in \Delta V} \varrho_i(\vec{r}, \theta) = \frac{1}{2\pi b}\sum_{i \in \Delta V} \delta_r\left(r_i'\right)\delta_\theta\left(\theta_i'\right) \tag{5.6}$$

$$\text{Coarse - grained flux: } \langle \varrho v\rangle(\vec{r},\theta):=\sum_{i\in\Delta V} v_i\, \varrho_i(\vec{r},\theta)=\frac{1}{2\pi b}\sum_{i\in\Delta V} v_i\delta_r\left(r_i'\right)\delta_\theta\left(\theta_i'\right) \quad (5.7)$$

$$\text{Coarse - grained rotational flux: } \langle \varrho\omega\rangle(\vec{r},\theta):=\sum_{i\in\Delta V} \omega_i\, \varrho_i(\vec{r},\theta)=\frac{1}{2\pi b}\sum_{i\in\Delta V} \omega_i\, \delta_r(r_i')\delta_\theta(\theta_i') \quad (5.8)$$

It should be noted that the summation here does not cancel out the SSDs, as would be the case for the GND $\vec{\rho}_G$ in equation (2.13). Say, a dislocation dipole would be represented by two different θ-space delta-like functions $\delta_\theta(\theta-\theta_i)$ at two θ_i values separated π apart, so that so the coarse-grained ϱ, $\langle \varrho v\rangle$, and $\langle \varrho\omega\rangle$ would not be zero.

Then, differentiating equation (5.6) with respect to time, we have

$$\dot{\varrho}(\vec{r},\theta)=-\underbrace{\sum_{i\in\Delta V}\left[\frac{\delta_\theta\left(\theta_i'\right)}{2\pi b}v_i\,\delta_r'\left(r_i'\right)\right]}_{\text{approach}}-\underbrace{\sum_{i\in\Delta V}\left[\frac{\delta_r\left(r_i'\right)}{2\pi b}\omega_i\,\delta_\theta'\left(\theta_i'\right)\right]}_{\text{tilting}} \quad (5.9)$$

From equation (5.7), we have

$$\frac{\partial\langle \varrho v\rangle}{\partial r_{\hat{\theta}}}=\frac{1}{2\pi b}\sum_{i\in\Delta V}\delta_\theta\left(\theta_i'\right)\frac{\partial\left[v_i\,\delta_r\left(r_i'\right)\right]}{\partial r_{\hat{\theta}}}=\frac{1}{2\pi b}\sum_{i\in\Delta V}\delta_\theta\left(\theta_i'\right)v_i\,\delta_r'\left(r_i'\right) \quad (5.10)$$

where $\partial r_{\hat{\theta}}$ refers to a movement of the field point $\vec{r}$ along the $\hat{\theta}$ direction for a particular dislocation character θ in the θ-space (see figure 5.2). The second step in equation (5.10) follows from the effect of the $\delta_\theta(\theta_i')$ selector, which guarantees that only those dislocations i with θ_i equal to the given θ value under consideration would participate in the summation. Then, for such dislocations, their velocities v_i approaching $\vec{r}$ would not change as $\vec{r}$ moves along the $\hat{\theta}$ direction, because $\hat{\theta}$ is just the direction of closest proximity to $\vec{r}$ for these dislocations.

Similarly, from equation (5.8), we have

$$\frac{\partial\langle \varrho\omega\rangle}{\partial\theta}=\sum_{i\in\Delta V}\left[\frac{\delta_r\left(r_i'\right)}{2\pi b}\omega_i\frac{\partial\delta_\theta(\theta_i')}{\partial\theta}\right]=\sum_{i\in\Delta V}\left[\frac{\delta_r\left(r_i'\right)}{2\pi b}\omega_i\,\delta_\theta'(\theta_i')\right] \quad (5.11)$$

since $\theta_i'=\theta-\theta_i$ and θ_i and ω_i are independent of a change $\partial\theta$ in the θ-domain.

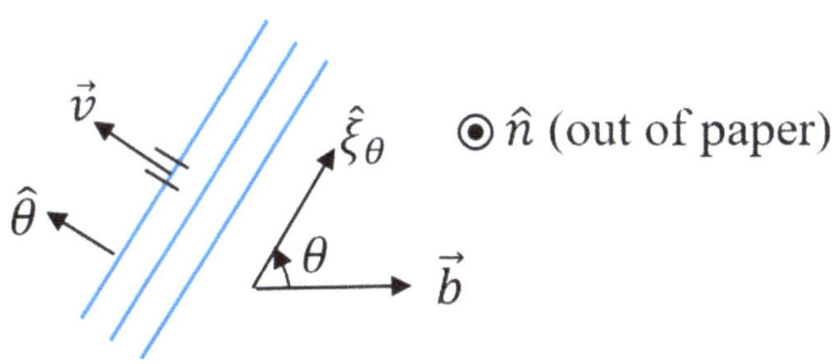

Figure 5.2. Diagram showing the definition of $\hat{\theta}$, $\hat{\xi}_\theta$ and $\hat{n}$, as well as $\vec{v}$.

Therefore, with equations (5.10) and (5.11), equation (5.9) becomes

$$\dot{\varrho}(\vec{r}, \theta) = -\underbrace{\frac{\partial\langle\varrho v\rangle}{\partial r_{\hat{\theta}}}}_{\text{approach}} - \underbrace{\frac{\partial\langle\varrho\omega\rangle}{\partial\theta}}_{\text{tilting}} \tag{5.12}$$

which is the kinematics law for the coarse-grained ADD for extensive microstructures.

Exercise 5.1. Why is it that ϱv qualifies for a flux by the latter's normal interpretation? What about $\varrho\omega$?

Solution: Consider a small rectangular area $\Delta_x \times \Delta_y$ threaded by the dislocation density ϱ moving at velocity v along the x direction, as shown in figure 5.3(a). Referring to the definition of ϱ in equation (2.14), the number of dislocations threading the area is $(\varrho d\theta)\Delta_x\Delta_y$ where $d\theta$ is a small interval in dislocation character. During time δt,

number of dislocations cutting across the length Δ_y $= \varrho d\theta \ \Delta_x\Delta_y \times \frac{v\delta t}{\Delta_x} = \varrho d\theta \ \Delta_y v\delta t$

number of dislocations per unit θ cutting across unit length per unit time $= \frac{\varrho d\theta \ \Delta_y v\delta t}{\Delta_y \delta t \ d\theta} = \varrho v$

Hence the flux ϱv is the number of dislocations, per unit character angle, cutting unit perpendicular length per unit time, in units of $\text{rad}^{-1}\ \text{m}^{-1}\ \text{s}^{-1}$.

For the rotational flux $\varrho\omega$, during time δt the density of dislocations (in #/m^2) rotating past a given dislocation character will be $\varrho\omega\delta t$ (figure 5.3(b)). Hence,

rotational flux $\varrho\omega =$ density of dislocations (#/m^2) rotating past a given dislocation character per unit time, in units of $\text{m}^{-2}\ \text{s}^{-1}$.

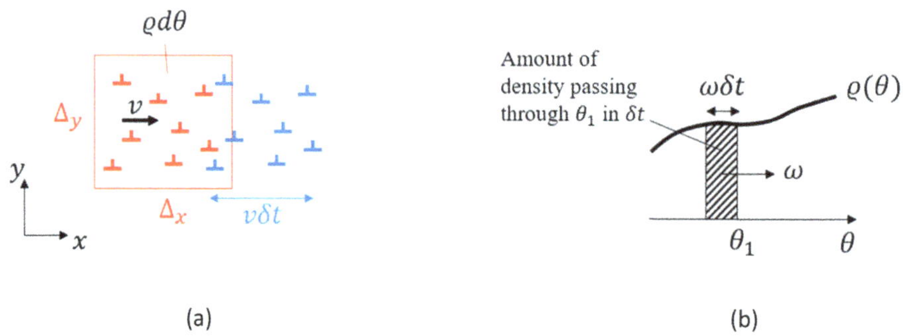

Figure 5.3. Dislocation fluxes: (a) ϱv, (b) $\varrho\omega$.

The units of ϱv and $\varrho\omega$ are different, but their gradients in the corresponding spaces that appear in equation (5.12), i.e. $\partial(\varrho v)/\partial r_{\hat{\theta}}$ and $\partial(\varrho\omega)/\partial\theta$, have the same unit of $m^{-2}\ rad^{-1}\ s^{-1}$, which is also the unit of $\dot{\varrho}$.

5.4 Alternative proofs of equation (5.12)

If a law is correct, there should be more than one way of proofing or interpreting it, and here, we show that this is the case for the kinematics law in equation (5.12).

5.4.1 Using Maxwell's law for parallel dislocation bundles

We have seen in chapter 3 that equation (3.9) holds for a bundle of parallel dislocations:

$$\dot{\vec{\rho}} = \underbrace{-[\vec{\nabla}(\rho v)\cdot\hat{v}]\,\hat{\xi}}_{\text{movement}} + \underbrace{\rho(\nabla_{\hat{\xi}}v)\hat{v}}_{\text{tilting}} \tag{3.9}$$

At a coarse-graining level pertinent to ADD, dislocations of different line directions and hence characters θ exist in each volume pixel, but equation (3.9) should be applicable to the specific density $\varrho(\theta)$ of each character θ in the entire population. Therefore, as the density $\varrho(\theta)$ of character θ moves in the $\vec{r}$-space and rotates in the θ-space, the rate of change in the vector-density $\vec{\rho}$ is given by:

$$\dot{\vec{\rho}}|_{\theta} = \underbrace{-\{\vec{\nabla}[\varrho(\theta)v]\cdot\hat{\theta}\}\,\hat{\xi}_{\theta}}_{\text{movement}} + \underbrace{\varrho(\theta)\omega\,\hat{\theta}}_{\text{tilting}} \tag{5.13}$$

where the notation $\dot{\vec{\rho}}|_{\theta}$ means that the $\dot{\vec{\rho}}$ is due to density $\varrho(\theta)$ of character θ, $\omega = \nabla_{\hat{\xi}_{\theta}}\ v$ is the rotational speed, and $\hat{\theta} = \hat{n}\times\hat{\xi}_{\theta}$ the direction on the slip-plane normal to the dislocations of character θ (figure 5.2). Since we are now looking at a particular dislocation character θ, $\hat{\theta}$ and $\hat{\xi}_{\theta}$ are unambiguously defined as shown in figure 5.2, and also, the velocity direction $\hat{v}$ is the same as the $\hat{\theta} = \hat{n}\times\hat{\xi}_{\theta}$ direction.

Now, consider the superposition of the $\dot{\vec{\rho}}|_{\theta}$ for different $\theta\in[0, 2\pi]$ from densities of different orientations in forming the change rate of ADD $\dot{\varrho}$. The two terms of movement and tilting in equation (5.13) will result in corresponding terms in the $\dot{\varrho}$ representation:

$$\dot{\varrho} = \dot{\varrho}^{\text{move}} + \dot{\varrho}^{\text{tilt}}, \tag{5.14}$$

where

$$\dot{\varrho}^{\text{move}} = -\ \vec{\nabla}\ (\varrho v)\cdot\hat{\theta} = -\frac{\partial(\varrho v)}{\partial r_{\hat{\theta}}} \tag{5.15}$$

with $r_{\hat{\theta}}$ being the coordinate along the $\hat{\theta}$ direction for each value of θ. Here, the partial differentiation $\partial(\varrho v)/\partial r_{\hat{\theta}}$ is carried out with respect to the spatial-domain variable $r_{\hat{\theta}}$ for each fixed value of θ in the dislocation character domain, so there is no ambiguity in the operation. For the tilting term in equation (5.14), just as ϱv is a flux in the spatial domain $\vec{r}$, the $\varrho\omega$ is a flux in the character domain θ. The term $\varrho\omega\hat{\theta}$

in equation (5.13) simply says that density $\varrho(\theta)$ is 'moving around' in the θ-space with flux $\varrho\omega$; thus if we treat θ as a linear coordinate and $\varrho(\theta)$ as objects moving along it with speed ω, then the corresponding conservative law (see equation (2.2)) will be:

$$\dot{\varrho}^{\text{tilt}} = -\frac{\partial(\varrho\omega)}{\partial\theta} \tag{5.16}$$

The kinematics law for $\varrho(\vec{r}, \theta)$ is thus shown to be:

$$\dot{\varrho} = \underbrace{-\vec{\nabla}(\varrho v)\cdot\hat{\theta}}_{\text{movement}} - \underbrace{\frac{\partial(\varrho\omega)}{\partial\theta}}_{\text{tilting}} \tag{5.17}$$

which is the same as that in equation (5.12).

Exercise 5.2. Prove equation (5.16).

Solution: The proof follows the standard way of proving equation (2.2), albeit now the domain concerned is the θ-space rather than the real space. As mentioned above, dislocations are rotating in the θ-space with flux $J_\theta = \varrho\omega$, while they are located at point $\vec{r}$ in the real space. From exercise 5.1,

> $J_\theta =$ density of dislocations (in $\#/\text{m}^2$) rotating past the dislocation character θ per unit time.

Consider an infinitesimal interval $\delta\theta$ at θ in the dislocation character space as shown in figure 5.4. The flux bringing dislocation density into the interval is J_θ while that bringing density out of the interval is $J_\theta + \delta J_\theta$. Then, during time δt,

$$\text{change in dislocation density (in } \#/m^2\text{) inside the interval at } = -\delta J_\theta\, \delta t \tag{5.18}$$

However, no dislocations of character θ are suddenly created or annihilated; the change in equation (5.18) above is simply due to dislocations rotating into or away from the character θ, so that it must also be equal to $\delta\varrho^{\text{tilt}}\,\delta\theta$, where $\delta\varrho^{\text{tilt}}$ is the change in $\varrho(\theta)$ (defined in equation (2.14)) due to dislocation rotation. Therefore,

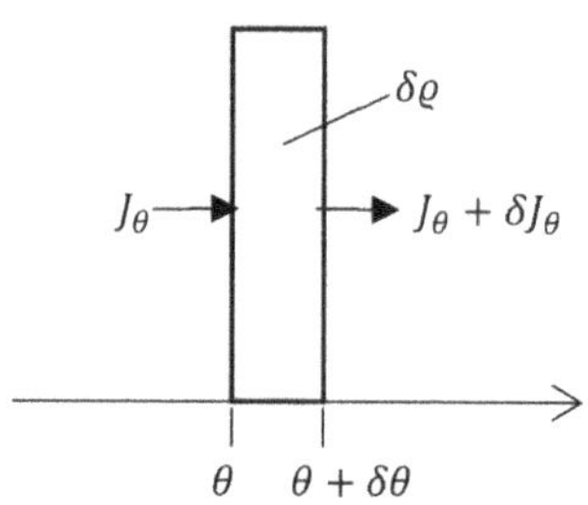

Figure 5.4. ϱ 'moving' in the dislocation character space with flux $J_\theta = \varrho\omega$, due to dislocations rotating while situated at point $\vec{r}$ in the real space.

$$\delta\varrho^{\text{tilt}}\,\delta\theta = -\delta J_\theta\,\delta t,$$

or

$$\dot{\varrho}^{\text{tilt}} = \frac{\delta\varrho^{\text{tilt}}}{\partial t} = -\frac{\partial J_\theta}{\partial\theta} = -\frac{\partial(\varrho\omega)}{\partial\theta}$$

Partial derivatives are used here because the context is for dislocations rotating while situated at a given spatial point $\vec{r}$, so θ is changing while $\vec{r}$ is held fixed.

5.4.2 Using Hochrainer's CDD

Based on Sedláček *et al*'s equation (3.11), Hochrainer *et al* (2007) also derived a kinematics law of the ADD but this is coupled to that of the dislocation curvature k according to:

$$\dot{\varrho} = -\vec{\nabla}\cdot(\varrho\vec{v}) - \frac{\partial(\varrho\omega)}{\partial\theta} + \varrho vk \tag{5.19}$$

We have seen in section 3.2.2, equations (3.10) and (3.11), that

$$\dot{\rho} = -\frac{\partial(\rho v)}{\partial r_{\hat{\theta}}} = -\vec{\nabla}\cdot(\rho\vec{v}) + \rho vk$$

for line density (#/m^2) with a particular dislocation character θ. Noting that $\rho = \varrho(\theta)d\theta$ for a particular interval $d\theta$, the above equation can be recast into:

$$\dot{\varrho}^{\text{move}} = -\frac{\partial(\varrho v)}{\partial r_{\hat{\theta}}} = -\vec{\nabla}\cdot(\varrho\vec{v}) + \varrho vk \tag{5.20}$$

With equation (5.20) and the common $\dot{\varrho}^{\text{tilt}}$ term $-\partial(\varrho\omega)/\partial\theta$ between equations (5.12) and (5.19), equation (5.19) is therefore fully compatible with equation (5.12). However, as noted also in section 3.2.2, there is no particular advantage of working with equation (5.19) since the curvature field $k(\vec{r}, \theta)$ is coupled.

5.5 Multiple slip and dynamics closure

Equation (5.12) is for the ADD on a specific slip system (figures 5.1 and 5.2), and when slip occurs on multiple systems, the ADD on each slip system has to be labelled and solved differently. Let α be the label of a particular slip system, and $\varrho_\alpha(\vec{r}, \theta)$ be the ADD on that system, so that, from equation (5.12),

$$\dot{\varrho}_\alpha(\vec{r}, \theta) = -\frac{\partial\langle\varrho_\alpha v_\alpha\rangle}{\partial r_{\hat{\theta}}} - \frac{\partial\langle\varrho_\alpha\omega_\alpha\rangle}{\partial\theta} \tag{5.21}$$

where quantities with subscript α refer to the slip system concerned. Thus, in solving multiple slip problems, equation (5.21) represents a system of simultaneous equations in which the velocity fields v_α are coupled between the slip systems via the dynamics laws.

For extensive microstructures, dynamics closure also makes use of the same velocity law as for intensive microstructures in equation (4.1), i.e.,

$$v_\alpha(\vec{r},\theta) = m_\alpha(\theta)\,\tau_\alpha^{\text{eff}}(\vec{r}) \tag{5.22}$$

However, as discussed in section 2.4 in chapter 2, the effective stress should now comprise additional terms such as:

$$\tau_\alpha^{\text{eff}}(\vec{r}) = \tau_\alpha^{\text{int}}(\vec{r}) + \tau_\alpha^{\text{app}} - \tau_p - \tau_\alpha^{\text{Taylor}}(\vec{r})$$

First, for extensive dislocation microstructures, dislocation core effects are not explicitly simulated, which means that the core-shielding factor f as per equation (4.24) is not applied. Therefore, a constant Peierls stress τ_{p} is explicitly included in the effective stress.

Another resistance to dislocation motion is the Taylor hardening stress due to the short-range interactions between dislocations, given by (Franciosi and Zaoui 1982, Leung *et al* 2015):

$$\tau_\alpha^{\text{Taylor}}(\vec{r}) = \mu b\sqrt{\sum_{\alpha'\neq\alpha} a_{\alpha\alpha'}\,\rho_{\text{total}}^{\alpha'}(\vec{r})} \tag{5.23A}$$

where

$$\rho_{\text{total}}^{\alpha'}(\vec{r}) = \int_0^{2\pi} \varrho_{\alpha'}(\vec{r},\theta)d\theta \tag{5.23B}$$

is the total dislocation length per unit volume for the interacting slip system α', and $a_{\alpha\alpha'}$ are interaction coefficients. Note that short-range interactions of the Taylor type are due to the collisions of dislocations on different slip systems and hence it is the total dislocation-length density, inclusive of any dipole contents present, that matters, rather than the GND density with the dipole contents subtracted out. The interaction coefficients $a_{\alpha\alpha'}$ are small fractions of unity, and their values depend on whether the interacting slip systems are coplanar (self hardening) or intersecting (latent hardening).

In the effective stress, τ_α^{int} is still the long-range elastic interactions between dislocation densities, but reconsideration of figure 4.3(b) would suggest that any dipole content at the source point $\vec{r}'$ would produce no net elastic interaction at the field point $\vec{r}$ because such interactions simply cancel out. Therefore, it is the GND density at the source point $\vec{r}'$, defined and calculated as

$$\vec{\rho}_G^{\alpha'}(\vec{r}') = \int_0^{2\pi} \varrho_{\alpha'}(\vec{r}',\theta)(\hat{s}\cos\theta + \hat{e}\sin\theta)d\theta \tag{5.24}$$

that produces the $\tau_\alpha^{\text{int}}(\vec{r})$ at the field point $\vec{r}$. In equation (5.24), $\hat{s}$ and $\hat{e}$ are unit vectors along the screw and edge directions of the α' slip system. Then, following equation (4.21A), τ_α^{int} is calculated as:

$$\tau_\alpha^{\text{int}}(\vec{r}) = \sum\nolimits_{\alpha'}\iiint g_{\alpha\alpha' h}(\vec{r}-\vec{r}')\xi_h^{\alpha'}\rho_{\text{G}}^{\alpha'}(\vec{r}')\,d^3r' \tag{5.25}$$

where $\rho_{\rm G}^{\alpha'}(\vec{r}') = |\vec{\rho}_{\rm G}^{\alpha'}(\vec{r}')|$, the $\xi_h^{\alpha'}$'s are the direction cosines of $\vec{\rho}_{\rm G}^{\alpha'}(\vec{r}')$ in the global coordinate system, and the kernel $g_{\alpha\alpha' h}$ is still that given in equation (4.21B), i.e.,

$$g_{\alpha\alpha' h}(\vec{r} - \vec{r}') = \frac{n_p^{\alpha} b_q^{\alpha} c_{pqrs}}{16\mu\pi(1-\nu)} \epsilon_{jsh} b_i^{\alpha'} c_{ijkl} \times \left[\frac{6\delta_{rk} a^2 (1-\nu)(x_l - x_l')}{R^5} + \frac{(3-4\nu)\delta_{rk}(x_l - x_l')}{R^3} + \frac{3(x_r - x_r')(x_k - x_k')(x_l - x_l')}{R^5} - \frac{\delta_{rl}(x_k - x_k') + \delta_{kl}(x_r - x_r')}{R^3} \right] \quad (4.21\text{B})$$

It should be noted that the sum $(\tau_{\rm p} + \tau_{\alpha}^{\rm Taylor}(\vec{r}))$ behaves as a friction called into play, i.e., its absolute value cannot exceed that of $(\tau_{\alpha}^{\rm int}(\vec{r}) + \tau_{\alpha}^{\rm app})$ so as to reverse the sign of $\tau^{\rm eff}(\vec{r})$. Therefore, the formal expression for $\tau_{\alpha}^{\rm eff}$ is:

$$\tau_{\alpha}^{\rm eff}(\vec{r}) = \begin{cases} \text{sign}\left(\tau_{\alpha}^{\rm int}(\vec{r}) + \tau_{\alpha}^{\rm app}\right) \left| \tau_{\alpha}^{\rm int}(\vec{r}) + \tau_{\alpha}^{\rm app} - \tau_p - \tau_{\alpha}^{\rm Taylor}(\vec{r}) \right|, & \text{if } \left| \tau_{\alpha}^{\rm int}(\vec{r}) + \tau_{\alpha}^{\rm app} \right| > \left| \tau_p + \tau_{\alpha}^{\rm Taylor}(\vec{r}) \right| \\ 0, & \text{otherwise} \end{cases} \quad (5.26)$$

Finally, it is also worth noting that the macroscopic plastic strain of the crystal is given by the Taylor–Orowan equation as:

$$\dot{\varepsilon}_p(\vec{r}) = \sum_{\alpha} (\vec{b}_{\alpha} \otimes \hat{n}_{\alpha}) \rho_G^{\alpha}(\vec{r}) v_{\alpha}(\vec{r}) \quad (5.27)$$

where $\rho_G^{\alpha} = |\vec{\rho}_G^{\alpha'}(\vec{r}')|$ as given by equation (5.24). Again, it is the GND density ρ_G^{α} that matters for the crystal strain because the movement of opposite-signed dislocations in a dipole would not produce any net strain of the crystal.

Exercise 5.3. Prove the Taylor–Orowan equation (5.27).

Solution:

Let us first consider the simple form of the Taylor–Orowan equation for a single slip system. Figure 5.5 shows that glide of different dislocations produces shear strain of the slip system. A dislocation i glided from one end of the crystal to the other will produce a shear displacement of the Burgers vector b; so if the dislocation

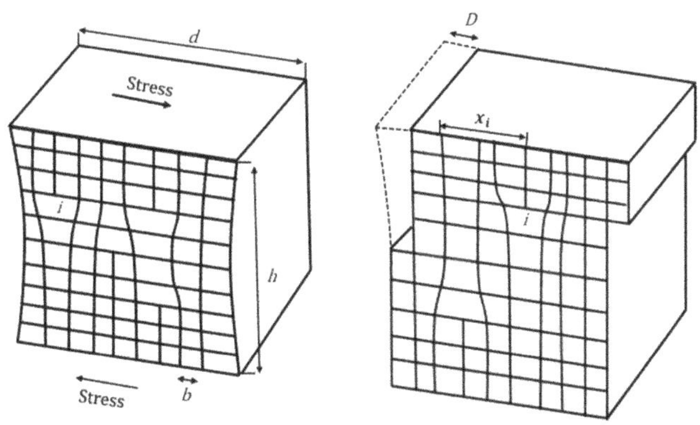

Figure 5.5. Dislocation movement in a single slip system producing strain.

only glides by a distance x_i instead of the whole crystal length d, the shear displacement will be $D = b(x_i/d)$, and hence the strain produced will be

$$\gamma_i = \frac{D}{h} = \left(\frac{x_i}{d}\right)\left(\frac{b}{h}\right),$$

where h is height of the crystal. The total shear strain produced by all dislocations in the crystal is therefore

$$\gamma = \sum_i \gamma_i = \frac{b}{hd}\sum_i x_i$$

and strain rate will be

$$\dot{\gamma} = \frac{d\gamma}{dt} = \frac{b}{hd}\sum_i \frac{dx_i}{dt} = \frac{bN\overline{v}}{dh} \tag{5.28}$$

where N is the number of mobile dislocations in the slip system, and $\overline{v} = \frac{1}{N}\sum_i \frac{dx_i}{dt}$ is their average velocity. As the dislocation density is the number of dislocations per threaded unit area, it is:

$$\rho = \frac{N}{hd} \tag{5.29}$$

Substituting equation (5.29) into equation (5.28), we have:

$$\dot{\gamma} = \rho b \overline{v} \tag{5.30}$$

Equation (5.30) gives the strain rate for the slip system (denoted by α) where the shear strain γ has displacement along the Burgers vector $\vec{b}_\alpha$ taking place over the slip plane with normal $\hat{n}_\alpha$. Moreover, it is obvious that the dislocation density ρ should be the GND density $\rho_{\mathrm{G}}^{\alpha}$, since movement of SSDs would not produce any net strain of the crystal. Thus, the strain-rate *tensor* should be:

$$\dot{\varepsilon}_{\mathrm{p}}^{\alpha} = \left(\hat{s}_\alpha \otimes \hat{n}_\alpha\right)\rho_{\mathrm{G}}^{\alpha} b_\alpha v_\alpha = \left(\vec{b}_\alpha \otimes \hat{n}_\alpha\right)\rho_{\mathrm{G}}^{\alpha} v_\alpha \tag{5.31}$$

where $\hat{s}_\alpha = \vec{b}_\alpha / b_\alpha$ is the slip direction of the slip system and v_α is the velocity of the GNDs. Expanding out in full,

$$\hat{s}_\alpha \otimes \hat{n}_\alpha = \begin{pmatrix} s_1 n_1 & s_1 n_2 & s_1 n_3 \\ s_2 n_1 & s_2 n_2 & s_2 n_3 \\ s_3 n_1 & s_3 n_2 & s_3 n_3 \end{pmatrix} \tag{5.32}$$

is known as the Schmid tensor, which has the property

$$\hat{s}_\alpha \otimes \hat{n}_\alpha \cdot \hat{e} = \begin{pmatrix} s_1 n_1 & s_1 n_2 & s_1 n_3 \\ s_2 n_1 & s_2 n_2 & s_2 n_3 \\ s_3 n_1 & s_3 n_2 & s_3 n_3 \end{pmatrix}\begin{pmatrix} e_1 \\ e_2 \\ e_3 \end{pmatrix} = \hat{s}_\alpha \left(\hat{n}_\alpha \cdot \hat{e}\right) \tag{5.33}$$

for any unit vector $\hat{e}$. The Schmid tensor therefore defines the slip system this way: its product with the normal $\hat{n}_\alpha$ of the slip plane is the shear direction $\hat{s}_\alpha$ itself, while that with any other plane perpendicular to the slip plane is zero. Note that the Schmid tensor is generally not symmetrical, i.e. $s_i n_j \neq s_j n_i$.

Finally, in a general, multi-slip situation, summation of equation (5.31) over the different slip systems α will give equation (5.27).

5.6 Elastoplastic framework for dynamics closure

The dynamics closure method in equation (5.25) makes use of the Mura's formula involving the elastic Green's function kernel $g_{\alpha\alpha' h}$ in equation (4.21B), which is valid only for infinite media. For general finite-sized media, dynamics closure can be carried out by FEM computation using the principles of elastoplastic framework, which is introduced here.

5.6.1 Basics of elastoplastic framework

While flow rules such as equation (5.27) define how the plastic strain rate can be calculated, the elastic deformation of the crystal should somehow be added upon to give the total deformation. In micromechanics, the scheme by which plastic and elastic deformations are combined is known as elastoplastic framework.

In micromechanics, deformation is represented by a deformation gradient **F** relating the deformed and undeformed configurations of the object under consideration. As shown in figure 5.6, the same material points are traced between the undeformed and deformed configurations during the object's deformation. Let

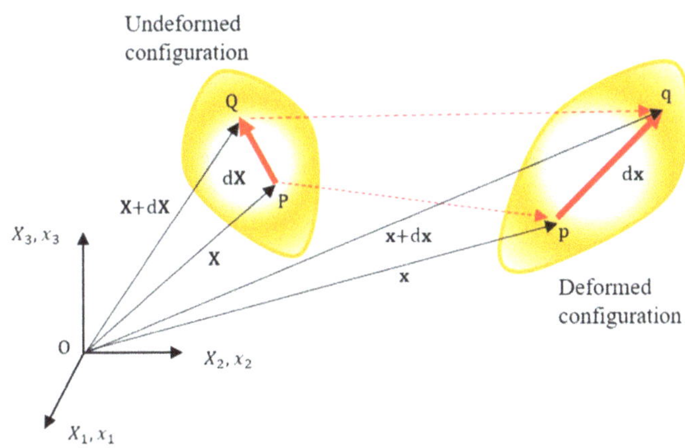

Figure 5.6. During deformation of the object, material point P in the undeformed configuration moves to p in the deformed configuration. In a fixed coordinate system in space, the position vector of P is $\boldsymbol{X}$ and that of p is $\boldsymbol{x}$. Similarly, a nearby point Q in the undeformed configuration is situated at $\boldsymbol{X}$+d**X**, and the corresponding point q in the deformed configuration is situated at $\boldsymbol{x}$+d**x**. The line segment $\overrightarrow{PQ}$ in the undeformed configuration is then deformed to $\overrightarrow{pq}$ in the deformed configuration.

$X = (X_1, X_2, X_3)$ denote the coordinates of a material point in the undeformed configuration,

$x = (x_1, x_2, x_3)$ denote the coordinates of the same material point in the deformed configuration,

and then the deformation of the object is defined by the function $x = x(X, t)$. The deformation gradient $\mathbf{F} = \{F_{ij}\}$ is then defined as a second-rank tensor as

$$F_{ij} = \frac{\partial x_i}{\partial X_j} \quad (i, j = 1, 2 \text{ or } 3) \tag{5.34}$$

Hence, $dx_j = \frac{\partial xj}{\partial X_1}dX_1 + \frac{\partial xj}{\partial X_2}dX_2 + \frac{\partial xj}{\partial X_3}dX_3 = \sum_{k=1}^{3} F_{jk}\, dX_k = F_{jk}\, dX_k$ (summation over repeated index is understood so summation sign is omitted), or

$$\mathrm{d}\mathbf{x} = \mathbf{F}\,\mathrm{d}\mathbf{X} \tag{5.35}$$

The time derivative of $\mathbf{F}$ is:

$$\dot{\mathbf{F}} = \frac{\partial \mathbf{F}}{\partial t} = \frac{\partial}{\partial t}\left[\frac{\partial \mathbf{x}(\mathbf{X}, t)}{\partial \mathbf{X}}\right] = \frac{\partial}{\partial \mathbf{X}}\left[\frac{\partial \mathbf{x}(\mathbf{X}, t)}{\partial t}\right] = \frac{\partial}{\partial \mathbf{X}}[\dot{\mathbf{x}}(\mathbf{X}, t)] = \frac{\partial}{\partial \mathbf{x}}[\dot{\mathbf{x}}(\mathbf{X}, t)] \cdot \frac{\partial \mathbf{x}(\mathbf{X}, t)}{\partial \mathbf{X}} = \boldsymbol{l}\,\mathbf{F}$$

or

$$\dot{F}_{ij} = l_{ik} F_{kj} \tag{5.36}$$

where

$$\boldsymbol{l} = \frac{\partial \dot{\mathbf{x}}}{\partial \mathbf{x}} \tag{5.37}$$

is defined as the spatial velocity gradient in the deformed coordinates $\mathbf{x}$. The velocity differential $\mathrm{d}\dot{\mathbf{x}}$ across a line segment $\mathrm{d}\mathbf{x}$ in the deformed configuration is therefore given as

$$\mathrm{d}\dot{\mathbf{x}} = \boldsymbol{l}\,\mathrm{d}\mathbf{x} \tag{5.38}$$

From equation (5.36),

$$\boldsymbol{l} = \dot{\mathbf{F}}\,\mathbf{F}^{-1},$$

or

$$l_{ij} = \frac{\partial v_i}{\partial x_j} = \dot{F}_{ik} {F^{-1}}_{kj} \tag{5.39}$$

As illustrated in figure 5.7, $\mathbf{F}$ is assumed to be decomposable into a plastic component $\mathbf{F}^{\mathrm{p}}$ and an elastic component $\mathbf{F}^{\mathrm{e}}$:

$$\mathbf{F} = \mathbf{F}^{\mathrm{e}}\,\mathbf{F}^{\mathrm{p}} \quad \text{or} \quad F_{ij} = F_{ik}^{e} F_{kj}^{p} \tag{5.40}$$

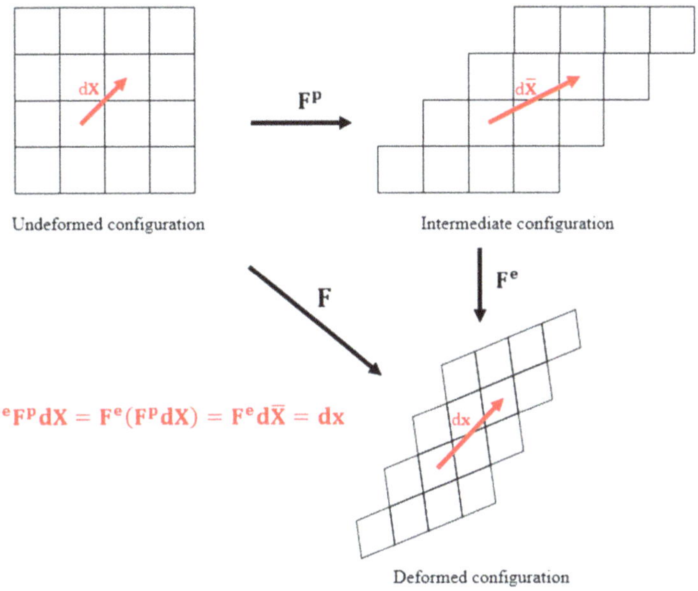

Figure 5.7. Decomposition of total deformation gradient $\mathbf{F}$ into elastic and plastic components $\mathbf{F^e}$ and $\mathbf{F^p}$. First, the reference (undeformed) configuration in which material coordinates are denoted as $\mathbf{X}$ is plastically deformed by $\mathbf{F^p}$ into the intermediate configuration, in which coordinates are denoted as $\bar{\mathbf{X}}$. In this case, a vector $d\mathbf{X}$ in the reference configuration is transformed into $d\bar{\mathbf{X}} = \mathbf{F^p}\, d\mathbf{X}$ in the intermediate configuration. Then, the intermediate configuration is elastically deformed into the final deformed configuration in which coordinates are denoted as $\mathbf{x}$. The $d\bar{\mathbf{X}}$ in the intermediate configuration is transformed into $d\mathbf{x} = \mathbf{F^e}\, d\bar{\mathbf{X}} = \mathbf{F^e}(\mathbf{F^p}\, d\mathbf{X}) = (\mathbf{F^eF^p})d\mathbf{X}$. The total transformation matrix is therefore $\mathbf{F} = \mathbf{F^eF^p}$.

where the transformations are defined by

$$d\mathbf{x} = \mathbf{F^e}d\bar{\mathbf{X}} \text{ and } d\bar{\mathbf{X}} = \mathbf{F^p}d\mathbf{X} \tag{5.41}$$

Then, from equation (5.38),

$$\begin{aligned} \boldsymbol{l} = \dot{\mathbf{F}}\,\mathbf{F}^{-1} &= \left[\frac{\partial}{\partial t}(\mathbf{F^e}\,\mathbf{F^p})\right](\mathbf{F^e}\,\mathbf{F^p})^{-1} = (\dot{\mathbf{F}}^{\mathbf{e}}\,\mathbf{F^p} + \mathbf{F^e}\,\dot{\mathbf{F}}^{\mathbf{p}})(\mathbf{F^{p-1}}\,\mathbf{F^{e-1}}) \\ &= \dot{\mathbf{F}}^{\mathbf{e}}\,\mathbf{F^{e-1}} + \mathbf{F^e}(\dot{\mathbf{F}}^{\mathbf{p}}\mathbf{F^{p-1}})\mathbf{F^{e-1}} = \dot{\mathbf{F}}^{\mathbf{e}}\,\mathbf{F^{e-1}} + \mathbf{F}^{e}\,\bar{\mathbf{L}}^{\mathbf{p}}\,\mathbf{F^{e-1}} \end{aligned} \tag{5.42}$$

where, in the last step, $\bar{\mathbf{L}}^{\mathbf{p}} = \dot{\mathbf{F}}^{\mathbf{p}}\mathbf{F^{p-1}}$ is the velocity gradient in the intermediate configuration that transforms any vector $d\bar{\mathbf{X}}$ in the intermediate configuration to the differential velocity $d\dot{\bar{\mathbf{X}}}$ across its end points (see equation (5.38) for $\boldsymbol{l}$ for the deformed configuration):

$$d\dot{\bar{\mathbf{X}}} = \bar{\mathbf{L}}^{\mathbf{p}}\, d\bar{\mathbf{X}} \tag{5.43}$$

Equation (5.42) may be rewritten as

$$\boldsymbol{l} = \boldsymbol{l}^{\mathrm{e}} + \boldsymbol{l}^{\mathrm{p}} \tag{5.44}$$

where

$$\boldsymbol{l}^{\mathrm{e}} = \dot{\mathbf{F}}^{\mathrm{e}}\,\mathbf{F}^{\mathrm{e}-1} \tag{5.45}$$

$$\boldsymbol{l}^{\mathrm{p}} = \mathbf{F}^{\mathrm{e}}\,\overline{\mathbf{L}}^{\mathrm{p}}\,\mathbf{F}^{\mathrm{e}-1}. \tag{5.46}$$

Thus, equation (5.42) or equation (5.44) indicates that the overall velocity gradient $\boldsymbol{l}$ is additively decomposable into an elastic part $\boldsymbol{l}^{\mathrm{e}}$ and a plastic part $\boldsymbol{l}^{\mathrm{p}}$.

Let us now explore the physical meanings of the three velocity gradients in equation (5.44):

(i) $\boldsymbol{l}$, as defined in equations (5.36)–(5.38), is the velocity gradient $(\partial\dot{\mathbf{x}}/\partial\mathbf{x})$ in the final deformed configuration deformed *elastoplastically* from the initial, undeformed configuration in figure 5.7, where the deformation is described by the total deformation gradient $\mathbf{F}$. $\boldsymbol{l}$ is unknown and the goal here is to compute it from the elastic and plastic velocity gradients in equation (5.44).

(ii) $\boldsymbol{l}^{\mathrm{e}}$ is given by $\boldsymbol{l}^{\mathrm{e}} = \dot{\mathbf{F}}^{\mathrm{e}}\,\mathbf{F}^{\mathrm{e}-1} = (\partial\dot{\mathbf{x}}/\partial\overline{\mathbf{X}})\cdot(\partial\overline{\mathbf{X}}/\partial\mathbf{x}) = \partial\dot{\mathbf{x}}/\partial\mathbf{x}$, which is also the velocity gradient in the deformed configuration in figure 5.7, but it must be remembered that this deformed configuration is obtained by *elastically* deforming the intermediate configuration in figure 5.7, where the deformation is described by the deformation gradient $\mathbf{F}^{\mathrm{e}}$. Thus, computationally, $\boldsymbol{l}^{\mathrm{e}}$ is calculated (by, say, FEM) as a *purely elasticity problem for the intermediate configuration* under the applied loading condition.

(iii) $\boldsymbol{l}^{\mathrm{p}}$ is given by $\boldsymbol{l}^{\mathrm{p}} = \mathbf{F}^{\mathrm{e}}\,\overline{\mathbf{L}}^{\mathrm{p}}\,\mathbf{F}^{\mathrm{e}-1} = (\partial\mathbf{x}/\partial\overline{\mathbf{X}})\cdot[(\partial\dot{\overline{\mathbf{X}}}/\partial\overline{\mathbf{X}})\cdot(\partial\overline{\mathbf{X}}/\partial\mathbf{x})] = (\partial\mathbf{x}/\partial\overline{\mathbf{X}})\cdot(\partial\dot{\overline{\mathbf{X}}}/\partial\mathbf{x})$, which is generally not equal to $(\partial\dot{\overline{\mathbf{X}}}/\partial\mathbf{x})\cdot(\partial\mathbf{x}/\partial\overline{\mathbf{X}}) = \partial\dot{\overline{\mathbf{X}}}/\partial\overline{\mathbf{X}} = \overline{\mathbf{L}}^{\mathrm{p}}$. However, the elastic deformation $\mathbf{F}^{\mathrm{e}} = \partial\mathbf{x}/\partial\overline{\mathbf{X}}$ is symmetric, and *if $\overline{\boldsymbol{L}}^{p} = \partial\dot{\overline{\boldsymbol{X}}}/\partial\overline{\boldsymbol{X}}$ is also symmetric*, then $(\partial\mathbf{x}/\partial\overline{\mathbf{X}})\cdot(\partial\dot{\overline{\mathbf{X}}}/\partial\mathbf{x})$ will be equal to $(\partial\dot{\overline{\mathbf{X}}}/\partial\mathbf{x})\cdot(\partial\mathbf{x}/\partial\overline{\mathbf{X}})$ so that $\boldsymbol{l}^{\mathrm{p}} = \overline{\mathbf{L}}^{\mathrm{p}}$, which is the velocity gradient in the intermediate configuration *plastically* deformed from the initial, undeformed configuration. Computationally, $\boldsymbol{l}^{\mathrm{p}}$ (or $\overline{\mathbf{L}}^{\mathrm{p}}$) can then be calculated as a *purely plasticity problem for the initial, undeformed configuration* under the applied loading condition.

Exercise 5.4. Show that if $\overline{\mathbf{L}}^{\mathrm{p}}$ is symmetric, then $\boldsymbol{l}^{\mathrm{p}} = \overline{\mathbf{L}}^{\mathrm{p}}$.

Solution: As shown above, $\boldsymbol{l}^{\mathrm{p}} = \mathbf{F}^{\mathrm{e}}\,\overline{\mathbf{L}}^{\mathrm{p}}\,\mathbf{F}^{\mathrm{e}-1} = (\partial\mathbf{x}/\partial\overline{\mathbf{X}})\cdot(\partial\dot{\overline{\mathbf{X}}}/\partial\mathbf{x})$. Here, $(\partial\mathbf{x}/\partial\overline{\mathbf{X}})$, etc., are 3×3 matrices, i.e. $(\partial\mathbf{x}/\partial\overline{\mathbf{X}}) = \{\partial x_i/\partial\overline{X}_j\}$, etc. Therefore, $\boldsymbol{l}^{\mathrm{p}} = \{l_{ij}^{p}\}$ where $l_{ij}^{p} = (\partial x_i/\partial\overline{X}_k)\cdot(\partial\dot{\overline{X}}_k/\partial x_j) \neq (\partial x_k/\partial\overline{X}_i)\cdot(\partial\dot{\overline{X}}_j/\partial x_k) = \partial\dot{\overline{X}}_j/\partial\overline{X}_i$ in general (summation implied over repeated index k). Alternatively, for two matrices $\boldsymbol{A}$ and $\boldsymbol{B}$ in general, $\boldsymbol{A}\cdot\boldsymbol{B} \neq \boldsymbol{B}\cdot\boldsymbol{A}$. However, $\boldsymbol{A} = \mathbf{F}^{\mathrm{e}} = \partial\mathbf{x}/\partial\overline{\mathbf{X}}$ here is symmetric as it is for an elastic deformation. If $\boldsymbol{B} = \partial\dot{\overline{\mathbf{X}}}/\partial\mathbf{x}$ is also symmetric, then $\boldsymbol{A}\cdot\boldsymbol{B} = \boldsymbol{B}\cdot\boldsymbol{A}$, and so $\boldsymbol{l}^{\mathrm{p}} = (\partial\dot{\overline{\mathbf{X}}}/\partial\mathbf{x})\cdot(\partial\mathbf{x}/\partial\overline{\mathbf{X}}) = \partial\dot{\overline{\mathbf{X}}}/\partial\overline{\mathbf{X}} = \overline{\mathbf{L}}^{\mathrm{p}}$. The question is therefore whether $\boldsymbol{B} = \partial\dot{\overline{\mathbf{X}}}/\partial\mathbf{x}$ is symmetric if $\overline{\mathbf{L}}^{\mathrm{p}}$ is symmetric.

Taking a step backward, we have $\boldsymbol{B} = \overline{\mathbf{L}}^{\mathrm{p}}\,\mathbf{F}^{\mathrm{e}-1}$, but since $\mathbf{F}^{\mathrm{e}}$ is symmetric, its inverse $\mathbf{F}^{\mathrm{e}-1}$ must also be symmetric. Therefore, if $\overline{\mathbf{L}}^{\mathrm{p}}$ is symmetric, $\boldsymbol{B}$ as the product of two symmetric matrices will be symmetric. The proof is therefore completed.

Exercise 5.5. Hashiguchi (2020) developed another decomposition rule for velocity gradients:

$$\overline{\mathbf{L}} = \overline{\mathbf{L}}^{\mathrm{e}} + \overline{\mathbf{L}}^{\mathrm{p}} \tag{5.47}$$

where

$$\overline{\mathbf{L}} = \mathbf{F}^{\mathrm{e}-1}\,\boldsymbol{l}\,\mathbf{F}^{\mathrm{e}}$$

$$\overline{\mathbf{L}}^{\mathrm{e}} = \mathbf{F}^{\mathrm{e}-1}\,\dot{\mathbf{F}}^{\mathrm{e}}$$

and $\overline{\mathbf{L}}^{\mathrm{p}}$ is that as before (i.e., defined in equation (5.43)). Show that if $\overline{\mathbf{L}}^{\mathrm{p}}$ is symmetric, then this decomposition leads to practically the same result as in equation (5.44).

Solution: First, since $\mathbf{F}^{\mathrm{e}}$ is symmetric, $\mathbf{F}^{\mathrm{e}-1}$ and $\dot{\mathbf{F}}^{\mathrm{e}}$ are also symmetric. Then in equation (5.47), $\overline{\mathbf{L}}^{\mathrm{e}} = \mathbf{F}^{\mathrm{e}-1}\,\dot{\mathbf{F}}^{\mathrm{e}} = \dot{\mathbf{F}}^{\mathrm{e}}\,\mathbf{F}^{\mathrm{e}-1}$, which is just $\boldsymbol{l}^{\mathrm{e}}$ as defined in equation (5.45). This is valid regardless of whether $\overline{\mathbf{L}}^{\mathrm{p}}$ is symmetric or not.

Secondly, we have seen from exercise 5.4 that if $\overline{\mathbf{L}}^{\mathrm{p}}$ is symmetric, then $\overline{\mathbf{L}}^{\mathrm{p}} = \boldsymbol{l}^{\mathrm{p}}$.

Thirdly, $\overline{\mathbf{L}} = \mathbf{F}^{\mathrm{e}-1}\,\boldsymbol{l}\,\mathbf{F}^{\mathrm{e}} = \mathbf{F}^{\mathrm{e}-1}\mathbf{F}^{\mathrm{e}}\,\boldsymbol{l} = \boldsymbol{l}$, since $\boldsymbol{l}$ is symmetric if $\overline{\mathbf{L}}^{\mathrm{p}}(= \boldsymbol{l}^{\mathrm{p}})$ is symmetric and $\boldsymbol{l}^{\mathrm{e}}$ must be symmetric.

Therefore, equation (5.47) is equivalent to equation (5.44) if $\overline{\mathbf{L}}^{\mathrm{p}}$ is symmetric.

5.6.2 Plastic velocity gradient and Schmid tensor

In the above subsection, the key result is equation (5.44), namely, the velocity gradient $\boldsymbol{l}$ in the deformed configuration of an object undergoing elastoplastic deformation, is the simple addition of an elastic component $\boldsymbol{l}^{\mathrm{e}}$ and a plastic component $\boldsymbol{l}^{\mathrm{p}}$. The elastic component $\boldsymbol{l}^{\mathrm{e}}$ can be computed as a well-defined pure elasticity problem starting from the intermediate (plastically deformed) configuration under the applied loading, and, as in CPFEM, this can be handled by FEM. For the plastic component $\boldsymbol{l}^{\mathrm{p}}$, this will be the simpler $\overline{\mathbf{L}}^{\mathrm{p}}$, the velocity gradient in the intermediate configuration plastically deformed from the initial configuration, if $\overline{\mathbf{L}}^{\mathrm{p}}$ is symmetric.

At this juncture, let us return to the Schmid tensor defined in equation (5.32), namely, $\hat{s}_\alpha \otimes \hat{n}_\alpha$ where $\hat{s}_\alpha =$ slip direction and $\hat{n}_\alpha =$ slip-plane normal. As discussed in equation (5.33), the Schmid tensor has the property that its product with any unit vector $\hat{e}$ gives $\hat{s}_\alpha$ multiplied by the direction cosine between $\hat{n}_\alpha$ and $\hat{e}$. Therefore, in the intermediate (plastically deforming) configuration shown in figure 5.8,

$$\left(\hat{s}_\alpha \otimes \hat{n}_\alpha\right) \cdot \mathrm{d}\mathbf{X} = \hat{s}_\alpha (\hat{n}_\alpha \cdot \hat{e})|\mathrm{d}\mathbf{X}| = \mathrm{d}h\ \hat{s}_\alpha \tag{5.48}$$

where $\hat{e}$ is the unit direction along $\mathrm{d}\mathbf{X}$, and $\mathrm{d}h = (\hat{n}_\alpha \cdot \hat{e})|\mathrm{d}\mathbf{X}|$ is the projected height of $\mathrm{d}\mathbf{X}$ along the slip-plane normal direction. The Schmid tensor $\hat{s}_\alpha \otimes \hat{n}_\alpha$ therefore

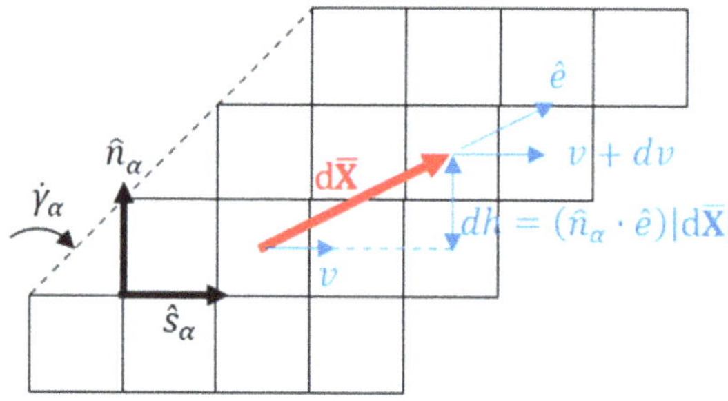

Figure 5.8. Velocity gradient in the intermediate (plastically deforming) configuration. Shear strain rate of the slip system ($\hat{s}_\alpha$, $\hat{n}_\alpha$) is $\dot{\gamma}_\alpha = \mathrm{d}v/\mathrm{d}h$.

transforms a line segment $\mathrm{d}\bar{\mathbf{X}}$ in the plastically deforming configuration into a vector of magnitude $\mathrm{d}h$ and direction $\hat{s}_\alpha$ which is the slip direction. Since the material velocities produced by slip must be along the $\hat{s}_\alpha$ direction, the velocity differential $\mathrm{d}\dot{\bar{\mathbf{X}}}$ across $\mathrm{d}\bar{\mathbf{X}}$ is

$$\mathrm{d}\dot{\bar{\mathbf{X}}} = \dot{\gamma}_\alpha \mathrm{d}h\, \hat{s}_\alpha = \dot{\gamma}_\alpha \left(\hat{s}_\alpha \otimes \hat{n}_\alpha\right) \cdot \mathrm{d}\bar{\mathbf{X}} \tag{5.49}$$

where $\dot{\gamma}_\alpha$ is the shear strain rate of the slip system and hence $\dot{\gamma}_\alpha \mathrm{d}h$ is the magnitude of $\mathrm{d}\dot{\bar{\mathbf{X}}}$. Comparing equation (5.49) with equation (5.43), the velocity gradient $\bar{\mathbf{L}}^\mathrm{p}$ takes the form of $\dot{\gamma}_\alpha\ \hat{s}_\alpha \otimes \hat{n}_\alpha$.

While $\bar{\mathbf{L}}^\mathrm{p}$ takes the form of $\dot{\gamma}_\alpha\ \hat{s}_\alpha \otimes \hat{n}_\alpha$, as stated in the text below equation (5.33), $\hat{s}_\alpha \otimes \hat{n}_\alpha$ is in general not symmetric, but as discussed above the elastoplastic framework will be greatly simplified if $\bar{\mathbf{L}}^\mathrm{p}$ is symmetric. Therefore, it is customary to symmetrise the Schmid tensor according to

$$\hat{s}_\alpha \otimes \hat{n}_\alpha \rightarrow \mathrm{sym}\left(\hat{s}_\alpha \otimes \hat{n}_\alpha\right) = \frac{1}{2}\left(\hat{s}_\alpha \otimes \hat{n}_\alpha + \hat{n}_\alpha \otimes \hat{s}_\alpha\right) \tag{5.50}$$

where the symmetrised Schmid tensor is also written as $\mathrm{sym}(\hat{s}_\alpha \otimes \hat{n}_\alpha)$. In constructing elastoplastic frameworks, the symmetrised Schmid tensor can replace the original one in simple CRSS-based flow rules such as equation (1.1):

$$\boldsymbol{l}^\mathrm{p} = \bar{\mathbf{L}}^\mathrm{p} = \sum_\alpha \dot{\gamma}_\alpha\ \mathrm{sym}\left(\hat{s}_\alpha \otimes \hat{n}_\alpha\right) \tag{5.51}$$

5.6.3 Overall and internal deformations in dislocation-based crystal plasticity

When the elastoplastic framework described above is applied to CRSS-based crystal plasticity models with flow rules of the type in equation (1.1), there is no ambiguity on the elastic and plastic deformation gradients $\mathbf{F}^\mathrm{e}$ and $\mathbf{F}^\mathrm{p}$ or the velocity gradients $\boldsymbol{l}^\mathrm{e}$ and $\boldsymbol{l}^\mathrm{p}$ of the crystal, since the internal dislocations, which are themselves elastic and plastic strain centres, are not considered. However, dislocation-based crystal

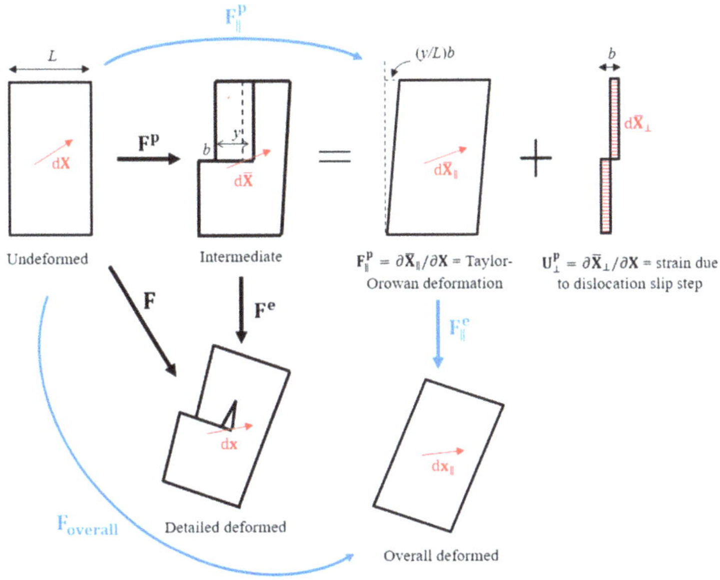

Figure 5.9. Decomposition of deformation into overall and internal components, the latter being the elastoplastic self fields of the internal dislocations. The overall deformation $\mathbf{F}_{\mathbf{overall}}$ is decomposable into a plastic component $\mathbf{F}^p_\parallel$ due to dislocation motion according to the Taylor–Orowan deformation, and an elastic component $\mathbf{F}^e_\parallel$ due to elastic deformation by external (not including internal) loading.

plasticity models including the present ADD-based model have the internal dislocations considered and so their strain fields would contribute to the deformation gradients. Fortunately, the strains of the internal dislocations are separable from the deformation gradients as follows (figure 5.9):

$$\mathbf{F}^p = \mathbf{F}^p_\parallel + \mathbf{U}^p_\perp \tag{5.52}$$

and

$$\mathbf{F}^e = \mathbf{F}^e_\parallel + \mathbf{U}^e_\perp \tag{5.53}$$

where
$\mathbf{U}^p_\perp$ = self plastic strain fields of the GNDs (those of SSDs cancel to zero at the coarse-grained level concerned),

$\mathbf{F}^p_\parallel$ = the coarse-grained, pure-shear deformation gradient due to the glide motion of GNDs without including their self strain fields and in accordance with the Taylor–Orowan equation (5.51) (movements of SSDs produce no net shear strain at the coarse-grained level concerned),

$\mathbf{U}^e_\perp$ = self elastic strain fields of the GNDs, and

$\mathbf{F}^e_\parallel$ = remaining part of the elastic deformation gradient of the crystal due to the applied loading.

Note that in the above, $\mathbf{F}_{\parallel}^{\mathrm{p}}$ and $\mathbf{F}_{\parallel}^{\mathrm{e}}$ are deformation gradients while $\mathbf{U}_{\perp}^{\mathrm{p}}$ and $\mathbf{U}_{\perp}^{\mathrm{e}}$ are strain tensors, i.e., if there are no deformation and no internal dislocations, then $\mathbf{F}_{\parallel}^{\mathrm{p}}$ and $\mathbf{F}_{\parallel}^{\mathrm{e}}$ equal $\mathbf{I}$ (identity tensor) while $\mathbf{U}_{\perp}^{\mathrm{p}}$ and $\mathbf{U}_{\perp}^{\mathrm{e}}$ equal $\mathbf{0}$ (zero tensor with all elements equal to zero). Also, $\mathbf{F}_{\parallel}^{\mathrm{p}}$ and $\mathbf{F}_{\parallel}^{\mathrm{e}}$ are *compatible* deformation gradient tensors satisfying $\vec{\nabla}_{\mathbf{X}} \times \mathbf{F}_{\parallel}^{\mathrm{p}} = \vec{\nabla}_{\overline{\mathbf{X}}} \times \mathbf{F}_{\parallel}^{\mathrm{e}} = \mathit{0}$, while $\mathbf{U}_{\perp}^{\mathrm{p}}$ and $\mathbf{U}_{\perp}^{\mathrm{e}}$ are *incompatible* deformation tensors obeying $\vec{\nabla}_{\mathbf{X}} \times \mathbf{U}_{\perp}^{\mathrm{p}} = -\vec{\nabla}_{\overline{\mathbf{X}}} \times \mathbf{U}_{\perp}^{\mathrm{e}} = \boldsymbol{\alpha}$, where $\boldsymbol{\alpha}$ is the Nye tensor defined in equation (2.10A) (exercise 5.6, see also Taupin *et al* (2008)).

Therefore, the deformation gradient can be decomposed into

$$\mathbf{F} = \mathbf{F}^{\mathrm{e}}\mathbf{F}^{\mathrm{p}} = \mathbf{F}_{\mathbf{overall}} + \mathbf{F}_{\mathbf{internal}} \tag{5.54}$$

where

$$\mathbf{F}_{\mathbf{overall}} = \mathbf{F}_{\parallel}^{\mathrm{e}}\,\mathbf{F}_{\parallel}^{\mathrm{p}} \tag{5.55}$$

and

$$\mathbf{F}_{\mathbf{internal}} = \mathbf{F}_{\parallel}^{\mathrm{e}}\mathbf{U}_{\perp}^{\mathrm{p}} + \mathbf{U}_{\perp}^{\mathrm{e}}\mathbf{F}_{\parallel}^{\mathrm{p}} + \mathbf{U}_{\perp}^{\mathrm{e}}\mathbf{U}_{\perp}^{\mathrm{p}}. \tag{5.56}$$

Here, $\mathbf{F}_{\mathbf{overall}} = \mathbf{F}_{\parallel}^{\mathrm{e}}\,\mathbf{F}_{\parallel}^{\mathrm{p}}$ describes the overall shape change of the crystal due to (i) the glide of the GNDs according to the Taylor–Orowan equation (5.51) and (ii) elastic deformation due to the applied loading (figure 5.9). $\mathbf{F}_{\mathbf{overall}}$ does not include the self fields of the internal dislocations, which are represented by $\mathbf{F}_{\mathbf{internal}}$. Formally, as illustrated in figure 5.9, we may write

$$\mathbf{F}^{\mathrm{p}} = \frac{\partial \overline{\mathbf{X}}}{\partial \mathbf{X}} = \frac{\partial(\overline{\mathbf{X}}_{\parallel} + \overline{\mathbf{X}}_{\perp})}{\partial \mathbf{X}} = \mathbf{F}_{\parallel}^{\mathrm{p}} + \mathbf{U}_{\perp}^{\mathrm{p}} \text{ and } \mathbf{F}^{\mathrm{e}} = \frac{\partial \mathbf{x}}{\partial \overline{\mathbf{X}}} = \frac{\partial(\mathbf{x}_{\parallel} + \mathbf{x}_{\perp})}{\partial \overline{\mathbf{X}}} = \mathbf{F}_{\parallel}^{\mathrm{e}} + \mathbf{U}_{\perp}^{\mathrm{e}}$$

where $\mathbf{U}_{\perp}^{\mathrm{p}} = (\partial\overline{\mathbf{X}}_{\perp}/\partial\mathbf{X})$, $\mathbf{U}_{\perp}^{\mathrm{e}} = (\partial\mathbf{x}_{\perp}/\partial\overline{\mathbf{X}})$, and

$$\left.\begin{aligned} \mathbf{F}_{\parallel}^{\mathrm{p}} &= \partial\overline{\mathbf{X}}_{\parallel}/\partial\mathbf{X} \\ \mathbf{F}_{\parallel}^{\mathrm{e}} &= \partial\mathbf{x}_{\parallel}/\partial\overline{\mathbf{X}} \end{aligned}\right\} \tag{5.57}$$

Here, an infinitesimal line segment $d\overline{\mathbf{X}}$ in the intermediate (plastically deformed) configuration is decomposed into $d\overline{\mathbf{X}} = d\overline{\mathbf{X}}_{\parallel} + d\overline{\mathbf{X}}_{\perp}$ where $d\overline{\mathbf{X}}_{\parallel} = \mathbf{F}_{\parallel}^{\mathrm{p}}\, d\mathbf{X}$ is the transformation of $d\mathbf{X}$ in the undeformed configuration due to the *average* plastic deformation as a result of GND movement given by the Taylor–Orowan equation, and $d\overline{\mathbf{X}}_{\perp}$ is the atomic-scale slip step in the intermediate configuration due to the self fields of the GNDs themselves (figure 5.9). Hence, $d\overline{\mathbf{X}}_{\perp}$ should be much shorter than any line segment $d\overline{\mathbf{X}}$ of the mesoscopic length scale on which dislocation densities are defined (see figure 2.7), i.e. $\partial\overline{\mathbf{X}}_{\perp}/\partial\overline{\mathbf{X}} \ll \mathbf{I}$ (the identity tensor). Therefore, $\partial\overline{\mathbf{X}}_{\parallel}/\partial\overline{\mathbf{X}} = \mathbf{I} - \partial\overline{\mathbf{X}}_{\perp}/\partial\overline{\mathbf{X}} \approx \mathbf{I}$, then $\mathbf{F}_{\parallel}^{\mathrm{e}} = (\partial\mathbf{x}_{\parallel}/\partial\overline{\mathbf{X}}) = (\partial\mathbf{x}_{\parallel}/\partial\overline{\mathbf{X}}_{\parallel}) \cdot (\partial\overline{\mathbf{X}}_{\parallel}/\partial\overline{\mathbf{X}}) \approx (\partial\mathbf{x}_{\parallel}/\partial\overline{\mathbf{X}}_{\parallel})$, and equations (5.55) and (5.57) can be rewritten as:

$$\left.\begin{aligned} \mathbf{F}_{\parallel}^{\mathrm{p}} &= \partial\overline{\mathbf{X}}_{\parallel}/\partial\mathbf{X} \\ \mathbf{F}_{\parallel}^{\mathrm{e}} &\approx \partial\mathbf{x}_{\parallel}/\partial\overline{\mathbf{X}}_{\parallel} \\ \mathbf{F}_{\mathbf{overall}} &= \mathbf{F}_{\parallel}^{\mathrm{e}}\,\mathbf{F}_{\parallel}^{\mathrm{p}} \approx \partial\mathbf{x}_{\parallel}/\partial\mathbf{X} \end{aligned}\right\} \tag{5.58}$$

where $d\mathbf{x}_{\parallel}$ is a line segment in the final elastoplastically deformed configuration that maps to $d\mathbf{X}$ in the initial undeformed configuration due to the *average* plastic deformation as a result of GND movement as well as the elastic deformation due to the applied loading. As mentioned above, the approximations in equation (5.58) involve the ignorance of the atomic-sized slip steps and self fields of the internal dislocations; the $\mathbf{F}_{\mathbf{overall}}$, $\mathbf{F}_{\parallel}^{\mathrm{e}}$ and $\mathbf{F}_{\parallel}^{\mathrm{p}}$ here are therefore analogous to the $\mathbf{F}$, $\mathbf{F}^{\mathrm{e}}$ and $\mathbf{F}^{\mathrm{p}}$, respectively, in CRSS-based crystal plasticity models in which internal dislocations are not considered. (It must be remembered, however, that the glide motion of the GNDs does depend on the mutual interactions of the self fields of the internal dislocations, which are considered in detail in the present ADD-based CDD scheme.)

Compared with the $\mathbf{F}_{\mathbf{internal}}$ in equation (5.56), which is the fluctuating deformation inside the crystal due to the self fields of the internal dislocations, the overall or *average* deformation $\mathbf{F}_{\mathbf{overall}} = \mathbf{F}_{\parallel}^{\mathrm{e}}\,\mathbf{F}_{\parallel}^{\mathrm{p}}$ under the applied loading may be of greater practical interest, since one is usually more interested in how the crystal changes shape macroscopically under external loading. Since equation (5.58) suggests that $\mathbf{F}_{\mathbf{overall}} = \mathbf{F}_{\parallel}^{\mathrm{e}}\,\mathbf{F}_{\parallel}^{\mathrm{p}}$ is still given by a two-step transformation $\mathbf{X}\xrightarrow{\mathbf{F}_{\parallel}^{\mathrm{p}}}\overline{\mathbf{X}}_{\parallel}\xrightarrow{\mathbf{F}_{\parallel}^{\mathrm{e}}}\mathbf{x}_{\parallel}$ as shown in figure 5.9, the properties in equations (5.41)–(5.46) are still applicable, in particular the additive decomposition of the velocity gradient:

$$\boldsymbol{l}_{\parallel} = \boldsymbol{l}_{\parallel}^{\mathrm{e}} + \boldsymbol{l}_{\parallel}^{\mathrm{p}} \tag{5.59}$$

where $\boldsymbol{l}_{\parallel} = \dot{\mathbf{F}}_{\parallel}\mathbf{F}_{\parallel}^{-1}$, $\boldsymbol{l}_{\parallel}^{\mathrm{e}} = \dot{\mathbf{F}}_{\parallel}^{\mathrm{e}}\mathbf{F}_{\parallel}^{\mathrm{e}-1}$ and $\boldsymbol{l}_{\parallel}^{\mathrm{p}} = \dot{\mathbf{F}}_{\parallel}^{\mathrm{p}}\mathbf{F}_{\parallel}^{\mathrm{p}-1}$. For ADD-based CDD, $\dot{\gamma}_{\alpha} = \rho_G^{\alpha} b_{\alpha} v_{\alpha}$ (see equations (5.27) and (5.51)), and by the definition of $\mathbf{F}_{\parallel}^{\mathrm{p}}$ the overall plastic strain rate should be set to be:

$$\boldsymbol{l}_{\parallel}^{\mathrm{p}} = \overline{\mathbf{L}}_{\parallel}^{\overline{\mathrm{p}}} = \sum_{\alpha} \rho_G^{\alpha} b_{\alpha} v_{\alpha}\ \mathrm{sym}\left(\hat{s}_{\alpha} \otimes \hat{n}_{\alpha}\right) \tag{5.60}$$

In general, there are two types of testing conditions, namely, load-controlled and strain-controlled:

- For load-controlled situations, the applied load is prescribed and the goal is to find the overall shape change of the crystal. If the applied stress is known, the elastic strain rate $\boldsymbol{l}_{\parallel}^{\mathrm{e}}$ is solvable from the applied stress rate using Hooke's Law, and then with $\boldsymbol{l}_{\parallel}^{\mathrm{p}}$ calculated from equation (5.60), the overall strain rate $\boldsymbol{l}_{\parallel}$ is obtainable from equation (5.59). More often, load rather than stress is prescribed and then solving for the elastic strain rate $\boldsymbol{l}_{\parallel}^{\mathrm{e}}$ from Hooke's law will require a numerical solver such as FEM.

- For strain-controlled testing, the overall strain rate $\boldsymbol{l}_{\parallel}$ is prescribed and the goal is to find the applied stress needed to achieve the shape change. With $\boldsymbol{l}_{\parallel}^{\mathrm{p}}$ calculated from equation (5.60), the elastic strain rate $\boldsymbol{l}_{\parallel}^{\mathrm{e}} = \boldsymbol{l}_{\parallel} - \boldsymbol{l}_{\parallel}^{\mathrm{p}}$ is known from equation (5.59), and then Hooke's law can be used to calculate the stress needed to achieve the shape change (see section 5.6.4).

Exercise 5.6. Show that $\boldsymbol{\alpha} = \vec{\nabla}_{\mathbf{X}} \times \mathbf{F}^{\mathrm{p}} = \vec{\nabla}_{\mathbf{X}} \times \mathbf{U}_{\perp}^{\mathrm{p}} = -\vec{\nabla}_{\overline{\mathbf{X}}} \times \mathbf{F}^{\mathrm{e}} = -\vec{\nabla}_{\overline{\mathbf{X}}} \times \mathbf{U}_{\perp}^{\mathrm{e}}$, and explain why $\vec{\nabla}_{\mathbf{X}} \times \mathbf{F}_{\parallel}^{\mathrm{p}} = \vec{\nabla}_{\overline{\mathbf{X}}} \times \mathbf{F}_{\parallel}^{\mathrm{e}} = \boldsymbol{0}$.

Solution: There are two (somewhat different but practically the same) definitions of the Burgers vector (Hirth and Lothe 1992). The more familiar is the 'true' Burgers vector defined in figure 5.10(a) by the FS/RH rule. Here a right-handed (RH) broken circuit with respect to the positive dislocation line direction (out of paper) is drawn from start (S) to finish (F) in the undeformed crystal (without the dislocation) while leaving a gap $\overrightarrow{\mathrm{FS}}$ which is then the Burgers vector $\vec{b}$, so that this circuit will be closed when mapped to the deformed (dislocated) crystal.

To prove the equations in the question, we use instead the other definition of $\vec{b}$, viz., the 'local' definition in figure 5.10(b) following the SF/RH rule (see also figure 1.2). Here, a right-handed (RH) broken circuit $\overline{C}$ (red contour in figure 5.10(b)) is drawn from start (S) to finish (F) in the plastically deformed (intermediate) configuration while leaving a cut $\overrightarrow{\mathrm{SF}}$, which is then the Burgers vector $\vec{b}$, so that this circuit will be a closed circuit C when mapped back into the undeformed crystal. Mathematically, this operation is represented by:

$$\vec{b} = \int_{\overline{C}} \mathrm{d}\overline{\mathbf{X}} = \oint_{C} \mathbf{F}^{\mathrm{p}} \cdot \mathrm{d}\mathbf{X} = \iint_{S} (\vec{\nabla}_{\mathbf{X}} \times \mathbf{F}^{\mathrm{p}}) \cdot \hat{\xi}\mathrm{d}S \tag{5.61}$$

where C is the Burgers circuit in the undeformed crystal, S a surface bounded by C and is therefore threaded by the dislocation, $\hat{\xi}$ the normal to S and is therefore along the dislocation line direction, and the second step is given by Stoke's theorem with the differential operators in the $\vec{\nabla}_{\mathbf{X}}$ based on the coordinates $\mathbf{X}$ in the undeformed configuration. On the other hand, from equation (2.10B), the Nye tensor has the property:

$$\iint_{S} \alpha \cdot \hat{\xi}\mathrm{d}S = \iint_{S} \{b_i \rho_j \xi_j\}\mathrm{d}S = \iint_{S} \{b_i \rho\}\mathrm{d}S = \{b_i\} \iint_{S} \rho \,\mathrm{d}S = \vec{b} \tag{5.62}$$

Comparing equations (5.61) and (5.62) and noting that the circuit C is arbitrary as long as it is encircling the dislocation, we have $\boldsymbol{\alpha} = \vec{\nabla}_{\mathbf{X}} \times \mathbf{F}^{\mathrm{p}}$. Furthermore, noting that $\mathbf{F}^{\mathrm{p}} = \mathbf{F}_{\parallel}^{\mathrm{p}} + \mathbf{U}_{\perp}^{\mathrm{p}}$, but $\vec{\nabla}_{\mathbf{X}} \times \mathbf{F}_{\parallel}^{\mathrm{p}} = \boldsymbol{0}$ since the deformation $\mathbf{F}_{\parallel}^{\mathrm{p}}$ does not arise from the self-field of any dislocation so that the corresponding Nye tensor is $\boldsymbol{0}$, we have $\boldsymbol{\alpha} = \vec{\nabla}_{\mathbf{X}} \times \mathbf{U}_{\perp}^{\mathrm{p}}$ proven.

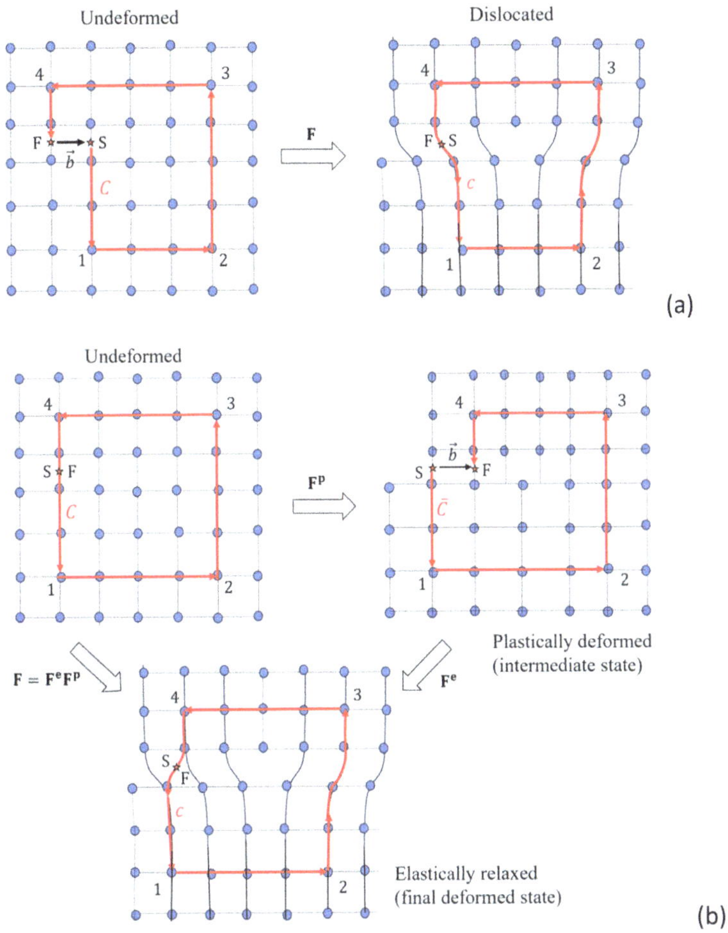

Figure 5.10. (a) Definition of the 'true' Burgers vector: a broken RH circuit with $\overrightarrow{\mathrm{FS}} = \vec{b}$ is first drawn in the undeformed crystal (left diagram) so that it becomes closed when mapped to the dislocated configuration (right diagram). (b) Definition of the 'local' Burgers vector (see also figure 1.2): a closed RH circuit is first drawn in the undeformed crystal (top left diagram), and then upon introduction of the dislocation by plastic deformation $\mathbf{F}^{\mathrm{p}}$, it becomes a broken contour with $\overrightarrow{\mathrm{SF}} = \vec{b}$ (top right diagram). Then, elastic relaxation of the plastically deformed dislocated crystal into the final, stable dislocated crystal by $\mathbf{F}^{\mathrm{e}}$ largely heals the previous gap $\overrightarrow{\mathrm{SF}} = \vec{b}$ in the former (bottom diagram). In all cases, positive dislocation line direction $\hat{\xi}$ is defined to point out of the paper.

In figure 5.10(b), there is also the $\mathbf{F}^{\mathrm{e}}$ deformation, which is elastic relaxation of the plastically deformed (intermediate) state into the final, stable dislocated state of the crystal. In this relaxation, the broken circuit $\overline{C}$ in the plastically deformed state will be mapped to a closed circuit c in the relaxed dislocated state with the previous gap $\overrightarrow{\mathrm{SF}} = \vec{b}$ closed, i.e.

$$\vec{0} = \oint_{c} \mathrm{d}\mathbf{x} = \int_{\overline{C}} \mathbf{F}^{\mathrm{e}}\mathrm{d}\overline{\mathbf{X}} = \oint_{\overline{C}+\overrightarrow{\mathrm{FS}}} \mathbf{F}^{\mathrm{e}}\mathrm{d}\overline{\mathbf{X}} - \mathbf{F}^{\mathrm{e}}\,\overrightarrow{\mathrm{FS}}$$

Here, $\int_{\overline{C}} \mathbf{F}^{\mathrm{e}} \mathrm{d}\overline{\mathbf{X}}$ is an integral along the broken circuit $\overline{C}$, and $\oint_{\overline{C}+\overrightarrow{\mathrm{FS}}} \mathbf{F}^{\mathrm{e}} \mathrm{d}\overline{\mathbf{X}}$ is an integral along the closed curve comprising $\overline{C}$ and $\overrightarrow{\mathrm{FS}}$. By Stoke's theorem, $\oint_{\overline{C}+\overrightarrow{\mathrm{FS}}} \mathbf{F}^{\mathrm{e}} \mathrm{d}\overline{\mathbf{X}} = \iint_{\overline{S}} (\vec{\nabla}_{\overline{\mathbf{X}}} \times \mathbf{F}^{\mathrm{e}}) \cdot \hat{\xi} \mathrm{d}\overline{S}$ where $\overline{S}$ is the surface bounded by $\overline{C} + \overrightarrow{\mathrm{FS}}$, and the $\vec{\nabla}_{\overline{\mathbf{X}}}$ operator is defined in the coordinates $\overline{\mathbf{X}}$ in the intermediate configuration. Therefore,

$$\iint_{\overline{S}} (\vec{\nabla}_{\overline{\mathbf{X}}} \times \mathbf{F}^{\mathrm{e}}) \cdot \hat{\xi} \mathrm{d}\overline{S} = \mathbf{F}^{\mathrm{e}}\, \overrightarrow{\mathrm{FS}} = -\mathbf{F}^{\mathrm{e}}\, \overrightarrow{\mathrm{SF}} = -\vec{b} \tag{5.63}$$

where $\vec{b} = \mathbf{F}^{\mathrm{e}}\, \overrightarrow{\mathrm{SF}}$, the $\overrightarrow{\mathrm{SF}}$ deformed by the elastic relaxation $\mathbf{F}^{\mathrm{e}}$, is still regarded as the Burgers vector. Comparing equation (5.63) with equation (5.62), we can then see that Nye tensor $\boldsymbol{\alpha} = -\vec{\nabla}_{\overline{\mathbf{X}}} \times \mathbf{F}^{\mathrm{e}}$, but since the deformation $\mathbf{F}^{\mathrm{e}}_{\parallel}$ involves no dislocation self-field, $\vec{\nabla}_{\overline{\mathbf{X}}} \times \mathbf{F}^{\mathrm{e}}_{\parallel} = \boldsymbol{0}$ so that $\boldsymbol{\alpha} = -\vec{\nabla}_{\overline{\mathbf{X}}} \times \mathbf{U}^{\mathrm{e}}_{\perp}$, thus completing the proof.

Exercise 5.7. There is yet another form for the Nye tensor: $\boldsymbol{\alpha} = -\mathrm{curl}\,(\mathbf{F}^{-1})$. Explain this.

Solution: Referring to figure 5.7, the operator 'curl' beginning with the lowercase 'c' means $(\vec{\nabla}_{\mathbf{x}}\times)$. Therefore,

$$\mathrm{curl}\,(\mathbf{F}^{-1}) = \vec{\nabla}_{\mathbf{x}} \times \mathbf{F}^{-1}$$

Here, we use the 'true' Burgers vector definition in figure 5.10(a), and consider the dislocated crystal as the final, elastoplastically deformed configuration. Considering the undeformed configuration, for an open curve C from S→1→2→3→4→F that will map into a closed curve c in the deformed state,

$$\vec{b} = \overrightarrow{\mathrm{FS}} = -\int_{C} \mathrm{d}\mathbf{X} = -\oint_{c} \mathbf{F}^{-1} \cdot \mathrm{d}\mathbf{x} = -\iint_{s} (\vec{\nabla}_{\mathbf{x}} \times \mathbf{F}^{-1}) \cdot \hat{\xi} \mathrm{d}s$$

where s a surface bounded by c and $\hat{\xi}$ its unit normal. Thus, comparing with the definition of the Nye tensor $\vec{b} = \iint_{s} \boldsymbol{\alpha} \cdot \hat{\xi} \mathrm{d}s$, $\boldsymbol{\alpha} = -\vec{\nabla}_{\mathbf{x}} \times \mathbf{F}^{-1}$.

5.6.4 Numerical implementation

To illustrate how the elastoplastic framework for ADD dynamics discussed so far can be implemented, we present a simple, straightforward numerical scheme for the strain-controlled situation, i.e., the overall strain-rate field $\boldsymbol{l}_{\parallel}$ of the crystal is specified and the goal is to solve for the externally applied stress field required to accomplish the imposed overall deformation. Suppose at the current time t, the overall elastic strain field $\boldsymbol{\varepsilon}_{\mathbf{e}(t)}$ and overall plastic strain field $\boldsymbol{\varepsilon}_{\mathbf{p}(t)}$ are already known (or, when $t = 0$, they are given as initial conditions). Again, as discussed in the previous section, $\boldsymbol{\varepsilon}_{\mathbf{e}(t)}$ and $\boldsymbol{\varepsilon}_{\mathbf{p}(t)}$ are overall elastic and plastic strains of a computational pixel excluding the strains of the internal dislocations; yet, they are the simulation parameters of each pixel and hence they vary between pixels. Since the total strain rate $\boldsymbol{l}_{\|}$ is specified, the total strain $\boldsymbol{\varepsilon}_{(t)}$, which is the sum of $\boldsymbol{\varepsilon}_{\mathbf{e}(t)}$ and $\boldsymbol{\varepsilon}_{\mathbf{p}(t)}$, is specified at time t; therefore, the elastic strain $\boldsymbol{\varepsilon}_{\mathbf{e}(t)} = \boldsymbol{\varepsilon}_{(t)} - \boldsymbol{\varepsilon}_{\mathbf{p}(t)}$ arises from the plastic strain $\boldsymbol{\varepsilon}_{\mathbf{p}(t)}$, the change of which is due to the density and movement of the GNDs inside the pixel, as well as the imposed total strain field $\boldsymbol{\varepsilon}_{(t)}$.

Stress in the current time t is calculated from Hooke's law as:

$$\boldsymbol{\sigma}_{(t)} = \mathbb{C}\colon \boldsymbol{\varepsilon}_{\mathbf{e}(t)} \tag{5.64}$$

where $\mathbb{C}$ is the elastic constant tensor. Resolved shear stresses on slip systems are calculated as (see exercise 5.7):

$$\tau_{\alpha}^{\text{int}} + \tau_{\alpha}^{\text{app}} = \boldsymbol{\sigma}_{(t)}\colon \text{sym}\left(\hat{s}_{\alpha} \otimes \hat{n}_{\alpha}\right) \tag{5.65}$$

Note that the $\boldsymbol{\sigma}_{(t)}$ here includes both the internal stress $\tau_{\alpha}^{\text{int}}$ due to the GNDs as well as the externally applied loading $\tau_{\alpha}^{\text{app}}$, since, as discussed above, the elastic strain $\boldsymbol{\varepsilon}_{\mathbf{e}(t)}$ includes the plastic strain $\boldsymbol{\varepsilon}_{\mathbf{p}(t)}$ due to the GNDs and the total shape change imposed by the external loading. As such, the method here avoids the use of the Mura's formula in equation (5.25), which is applicable to infinite media only.

ADD dynamics is then solved by using equation (5.65) for $(\tau_{\alpha}^{\text{int}} + \tau_{\alpha}^{\text{app}})$ in the effective glide stress $\tau_{\alpha}^{\text{eff}}$ in equation (5.26), to yield the $\rho_{G}^{\alpha}(\vec{r})$ and $v_{\alpha}(\vec{r})$ fields, and then $\boldsymbol{l}_{\|}^{\mathbf{p}}$ is calculated from equation (5.60). During the next time step Δt, the plastic strain increment is calculated as:

$$\Delta\boldsymbol{\varepsilon}_{\mathbf{p}} = \boldsymbol{l}_{\|}^{\mathbf{p}}\, \Delta t \tag{5.66}$$

The increment in true strain is given from the imposed $\boldsymbol{l}_{\|}$ as:

$$\Delta\boldsymbol{\varepsilon} = \boldsymbol{l}_{\|}\, \Delta t \tag{5.67}$$

and from equation (5.59), $\boldsymbol{l}_{\|}$ is additively decomposed into elastic and plastic components; hence, strains are also additively superimposed. Therefore, $\Delta\boldsymbol{\varepsilon} = \Delta\boldsymbol{\varepsilon}_{\mathbf{e}} + \Delta\boldsymbol{\varepsilon}_{\mathbf{p}}$, and hence the elastic strain at the next time step $t + \Delta t$ is:

$$\boldsymbol{\varepsilon}_{\mathbf{e}(t+\Delta t)} = \boldsymbol{\varepsilon}_{\mathbf{e}(t)} + \Delta\boldsymbol{\varepsilon} - \Delta\boldsymbol{\varepsilon}_{\mathbf{p}} \tag{5.68}$$

where $\boldsymbol{\varepsilon}_{\mathbf{e}(t+\Delta t)} - \boldsymbol{\varepsilon}_{\mathbf{e}(t)} = \Delta\boldsymbol{\varepsilon}_{\mathbf{e}}$ is the increment in the elastic strain during the time step. Then, the process loops back to equation (5.64) to calculate the stress $\boldsymbol{\sigma}_{(t+\Delta t)}$ in the next cycle.

Although, as mentioned above, the method here avoids the use of Mura's formula and is therefore generally applicable to finite-sized media, in usual circumstances the entire strain-rate field $\boldsymbol{l}_{\|}$ is not imposed but rather, the boundary of the medium is prescribed. Then, the elastic strain $\boldsymbol{\varepsilon}_{\mathbf{e}}$ will need to be solved by rather expensive computation by the finite-element method (FEM) from the imposed boundary conditions, via consideration of mechanical equilibrium or elastodynamics.

Exercise 5.8. Interpret equations (5.64) and (5.65).

Solution: Both equations of $\boldsymbol{\sigma} = \mathbb{C}: \boldsymbol{\varepsilon}_\mathbf{e}$ and $\tau_\alpha^{\mathrm{app}} = \boldsymbol{\sigma}: \mathrm{sym}(\hat{s}_\alpha \otimes \hat{n}_\alpha)$ involve the double dot product ':' but its meaning in the two equations is not the same.

In the Hooke's Law $\boldsymbol{\sigma} = \mathbb{C}: \boldsymbol{\varepsilon}_\mathbf{e}$, the double dot product is between a fourth-order tensor $\mathbb{C} = \{c_{ijkl}\}$ where the c_{ijkl}'s $(i, j, k, l = 1, 2, 3)$ are elastic constants, and a second-order tensor $\boldsymbol{\varepsilon}_\mathbf{e} = \{\varepsilon_{ekl}\}$ where the ε_{ekl}'s $(k, l = 1, 2, 3)$ are elastic strain components. Then, the double dot product is a second-order tensor of $=\{\sigma_{ij}\}$, where

$$\sigma_{ij} = c_{ijkl}\varepsilon_{ekl}(\text{summation over repeated indices } k \text{ and } l) \tag{5.69}$$

For isotropically elastic materials, the elastic constants c_{ijkl}'s are given by equation (4.22).

In $\tau_\alpha^{\mathrm{app}} = \boldsymbol{\sigma}: \mathrm{sym}(\hat{s}_\alpha \otimes \hat{n}_\alpha)$, the double dot product is now between two second-order tensors $\boldsymbol{\sigma} = \{\sigma_{ij}\}$ and $\mathrm{sym}(\hat{s}_\alpha \otimes \hat{n}_\alpha)$. In general, for two second-order tensors $\mathbf{A}$ and $\mathbf{B}$, their double dot product is a *scalar* defined as $\mathbf{A}: \mathbf{B} = A_{ij}B_{ij}$ (summation over i and j). Hence,

$$\tau_\alpha^{\mathrm{app}} = \boldsymbol{\sigma}: \mathrm{sym}\left(\hat{s}_\alpha \otimes \hat{n}_\alpha\right) = \frac{1}{2}\sigma_{ij}(s_i n_j + s_j n_i) \text{ (summation over } i \text{ and } j) \tag{5.70}$$

References

Acharya A and Roy A 2006 Size effects and idealized dislocation microstructure at small scales: predictions of a phenomenological model of mesoscopic field dislocation mechanics: part I *J. Mech. Phys. Solids* **54** 1687

Franciosi P and Zaoui A 1982 Multislip in F.C.C. crystals: a theoretical approach compared with experimental data *Acta Metall.* **30** 1627–37

Hashiguchi K 2020 *Nonlinear Continuum Mechanics for Finite Elasticity-Plasticity: Multiplicative Decomposition with Subloading Surface Model* (Amsterdam: Elsevier)

Hirth J P and Lothe J 1992 *Theory of Dislocations* 2nd edn (Malabar, FL: Krieger Publishing)

Hochrainer T, Zaiser M and Gumbsch P 2007 A three-dimensional continuum theory of dislocation systems: kinematics and mean-field formulation *Philos. Mag.* **87** 1261

Leung H S, Leung P S S, Cheng B and Ngan A H W 2015 A new dislocation-density-function dynamics scheme for computational crystal plasticity by explicit consideration of dislocation elastic interactions *Int. J. Plast.* **67** 1–25

Ngan A H W 2017 Dislocation-density kinematics: a simple evolution equation for dislocation density involving movement and tilting of dislocations *MRS Commun.* **7** 583–90

Taupin V, Varadhan S, Fressengeas C and Beaudoin A J 2008 Directionality of yield point in strain-aged steels: the role of polar dislocations *Acta Mater.* **56** 3002–10

Chapter 6

Numerical scheme for ADD simulation

6.1 Realistic ADD for connected-line microstructures

As mentioned previously, the evolution law in equation (5.12) for ADD is an exact kinematics law that preserves the connectivity of the dislocation lines. Therefore, the first task of carrying out ADD simulation is to generate an initial $\varrho(\vec{r}, \theta)$ distribution that is physically correct, namely, one that matches a microstructure of connected lines. A randomly generated density distribution that does not represent connected lines would not be physically realistic or commensurate with the kinematics law in equation (5.12). To guarantee that a density distribution corresponds to a microstructure of connected lines, one may propose the following procedure:

(i) Set up a microstructure of discrete curves on the slip plane, as shown in the example in figure 6.1(a).

(ii) Then, for each discrete curve i in the picture, apply the concept in figure 5.1 and equation (5.3) to set up its ADD:

$$\varrho_i(\vec{r}, \theta) = \frac{\delta_r(|\vec{r} - \vec{r}_i|)}{b}\frac{\delta_\theta(\theta - \theta_i)}{2\pi} \tag{6.1}$$

where $\vec{r}_i$ is the point on the curve that is closest to the field point $\vec{r}$, and θ_i is the dislocation character (or inclination angle of the tangent to the curve) at $\vec{r}_i$ (see figure 5.1).

(iii) Then, the total ADD is obtained by summing over the curves:

$$\varrho(\vec{r}, \theta) = \sum_i \varrho_i(\vec{r}, \theta) = \sum_i \frac{\delta_r(|\vec{r} - \vec{r}_i|)}{b}\frac{\delta_\theta(\theta - \theta_i)}{2\pi} \tag{6.2}$$

As an illustration, the functions δ_r and δ_θ may be chosen to be Gaussian functions, hence, equation (6.1) becomes:

$$\varrho(\vec{r}, \theta) = \frac{1}{4\pi^2\sigma\sigma' b}\sum_{i=1}^{N} \exp\left[-\frac{1}{2}\left(\frac{|\vec{r} - \vec{r}_i|}{\sigma}\right)^2\right]\exp\left[-\frac{1}{2}\left(\frac{\theta - \theta_i}{\sigma'}\right)^2\right] \tag{6.3}$$

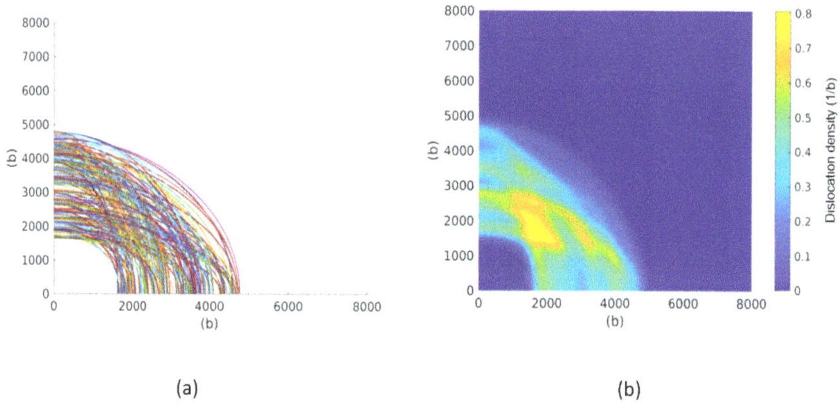

Figure 6.1. Setting up initial ADD for connected lines. (a) A large number (300 here) of discrete elliptic curves set up on a slip plane. (b) The corresponding coarse-grained density map by applying equation (6.3) to the discrete-line picture in (a).

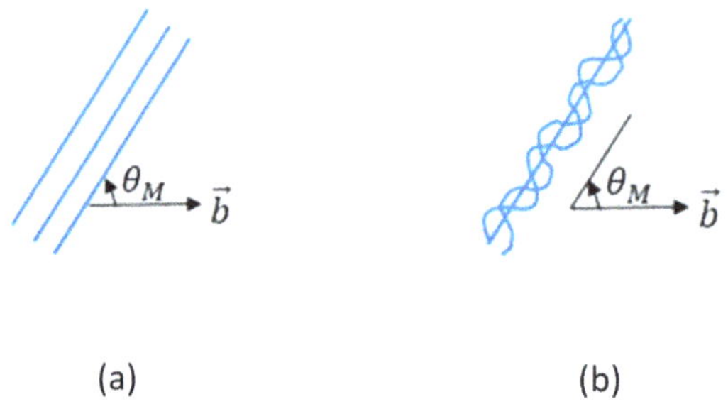

Figure 6.2. Coarse-grained line contents (A) with only spread in spatial space, (B) with spread in both spatial and θ space.

where σ and σ' are, respectively, the standard deviation for the density in the spatial and θ domains. Figure 6.1(b) shows the coarse-grained density map $\rho(\vec{r}) = \int_0^{2\pi} \varrho(\vec{r}, \theta) d\theta$ by applying equation (6.3) to the 300 discrete lines in figure 6.1(a).

While figure 6.1 may look impressive, let us not celebrate so early! When it comes to numerical implementation, there are complications arising from the δ_θ function in equation (6.1) that we have to consider carefully.

The prime objective of introducing the δ_θ function with some spread in ϱ_i is of course to avoid singularities in handling the derivative $\partial\langle\varrho\omega\rangle/\partial\theta$ in the tilting term in equation (5.12). But in doing so, we also introduce extra line contents with orientations different from the modal value θ_M which is the orientation θ_i of the original line in figure 5.1. Figure 6.2 shows the difference.

To understand the behavior of the added extra line contents with orientations different from the original θ_M, consider that the original line has some local curvature as shown in figure 6.3(a). In terms of a cylindrical coordinate system

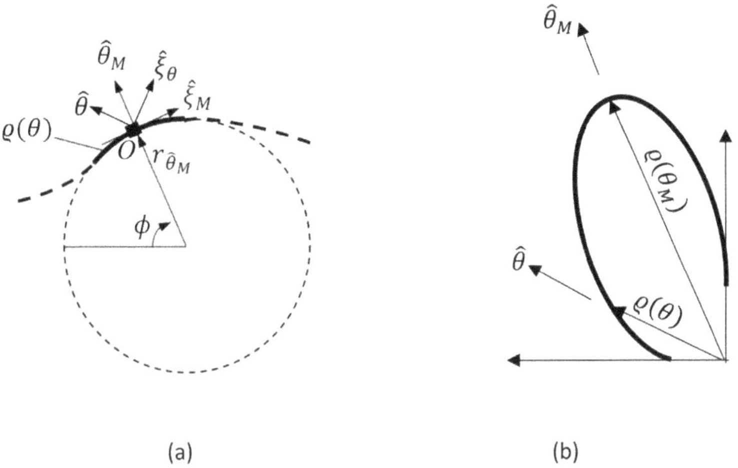

Figure 6.3. (a) A segment of a curved line coarse-grained with a spread in the θ-space, superimposed on a cylindrical coordinate system ($r_{\hat{\theta}_M}$, ϕ) with the origin at the centre of curvature of the line. (b) Polar plot of the ADD $\varrho(\theta)$ at point O in (a), with θ_M as the main (modal) line orientation. (Reproduced from Kalaei *et al* (2022), copyright (2022) with permission from Elsevier.)

($r_{\hat{\theta}_M}$, ϕ) with the origin placed at the centre of curvature, the gradient operator (see section 3.2.2) is:

$$\vec{\nabla}_M = \hat{\theta}_M \frac{\partial}{\partial r_{\hat{\theta}_M}} + \hat{\xi}_M \frac{1}{r_{\hat{\theta}_M}} \frac{\partial}{\partial \phi} \tag{6.4}$$

The movement term of $\dot{\varrho}$ in equation (5.15) is then:

$$\dot{\varrho}^{\text{move}}(\theta) = -[\vec{\nabla}_M(\varrho(\theta)v)] \cdot \hat{\theta} = -\left[(\hat{\theta}_M \cdot \hat{\theta})\frac{\partial \varrho(\theta)v}{\partial r_{\hat{\theta}_M}} + (\hat{\xi}_M \cdot \hat{\theta})\frac{1}{r_{\hat{\theta}_M}}\frac{\partial \varrho(\theta)v}{\partial \phi}\right] \tag{6.5}$$

However, the requirement of line connectivity is the divergence-free condition $\vec{\nabla}_M\ \varrho(\theta) \cdot \hat{\xi}_\theta = 0$, explained in exercise 3.2, where $\hat{\xi}_\theta$ is the line direction for a general θ that may or may not equal θ_M (see figure 6.3(A)). Therefore,

$$\vec{\nabla}_M\ \varrho(\theta) \cdot \hat{\xi}_\theta = (\hat{\theta}_M \cdot \hat{\xi}_\theta)\frac{\partial \varrho(\theta)}{\partial r_{\hat{\theta}_M}} + (\hat{\xi}_M \cdot \hat{\xi}_\theta)\frac{1}{r_{\hat{\theta}_M}}\frac{\partial \varrho(\theta)}{\partial \phi} = 0 \tag{6.6}$$

Eliminating $\frac{1}{r_{\hat{\theta}_M}}\frac{\partial \varrho(\theta)}{\partial \phi}$ between equations (6.5) and (6.6) yields

$$\dot{\varrho}^{\text{move}}(\theta) = -\left[(\hat{\theta}_M \cdot \hat{\theta}) - \frac{(\hat{\xi}_M \cdot \hat{\theta})(\hat{\theta}_M \cdot \hat{\xi}_\theta)}{(\hat{\xi}_M \cdot \hat{\xi}_\theta)}\right]\frac{\partial \varrho(\theta)v}{\partial r_{\hat{\theta}_M}},$$

but from the geometry shown in figure 6.3(a), $(\hat{\theta}_M \cdot \hat{\theta}) = (\hat{\xi}_M \cdot \hat{\xi}_\theta)$, $(\hat{\xi}_M \cdot \hat{\theta}) = -(\hat{\theta}_M \cdot \hat{\xi}_\theta)$, and $(\hat{\theta}_M \cdot \hat{\theta})^2 + (\hat{\theta}_M \cdot \hat{\xi}_\theta)^2 = 1$, so that

$$\left[(\hat{\theta}_M \cdot \hat{\theta}) - \frac{(\hat{\xi}_M \cdot \hat{\theta})(\hat{\theta}_M \cdot \hat{\xi}_\theta)}{(\hat{\xi}_M \cdot \hat{\xi}_\theta)}\right] = \frac{1}{(\hat{\theta}_M \cdot \hat{\theta})}.$$

Hence,

$$\dot{\varrho}^{\text{move}}(\theta) = -\frac{\partial}{\partial r_{\hat{\theta}_M}}[\varrho(\theta)(v/(\hat{\theta}_M \cdot \hat{\theta}))] \tag{6.7}$$

Equation (6.7) states that for any θ, $\varrho(\theta)$ evolves conservatively along the normal direction $\hat{\theta}_M$ of the modal line orientation, with an effective velocity $v/(\hat{\theta}_M \cdot \hat{\theta})$ that depends on the individual orientation θ. Therefore, if we specify the original line with orientation θ_M to move with velocity v, but we introduce extra densities with orientations $\theta \neq \theta_M$ by using a δ_θ function with a finite spread, then the extra densities will move with velocities $v/(\hat{\theta}_M \cdot \hat{\theta})$ that are higher than the velocity v of the modal orientation θ_M. This means that the coarse-grained line will widen as it moves.

So, why don't we make the δ_θ function a Kronecker delta function with infinitesimal spread? In this case, no extra line contents with unwanted orientations will be added, but then an implementing numerical scheme would be very challenging. Apart from encountering singularities in handling the tilting term in equation (5.12), for a discretization scheme as illustrated in figure 6.4, the $\varrho(\theta)|_O$ at the centre of a pixel ABCD should be averaged from the $\varrho(\theta)|_A$, $\varrho(\theta)|_B$, $\varrho(\theta)|_C$ and $\varrho(\theta)|_D$ in the neighbourhood, but then if the δ_θ is a sharp spike, the $\varrho(\theta)|_{A \text{ to } D}$ would all be close to zero at the modal value $\theta = \theta_M|_O$ for the density at the pixel centre O. We thus end up with the dilemma that for any discretization scheme in the θ-space to be usable, there should be spread in the δ_θ function, but as discussed above, this will introduce unwanted line orientations that will misbehave.

Are we stuck? Fortunately, no, and in fact, the problem above can be fixed easily by a simple modification to the numerical scheme for the kinematics law. Recalling the slip-plane discretization scheme in figure 3.6, and the derivation leading to equation (3.16), the finite-difference equation for $\dot{\varrho}^{\text{move}}$ may seem to be:

$$\dot{\varrho}^{\text{move}}(\theta)|_{ABCD} \ \Delta s = -[\Delta(\varrho v(\theta))|_{\overline{AB}} \ \hat{s} + \Delta(\varrho v(\theta))|_{\overline{CD}} \ \hat{e}] \cdot \hat{\theta} \tag{6.8}$$

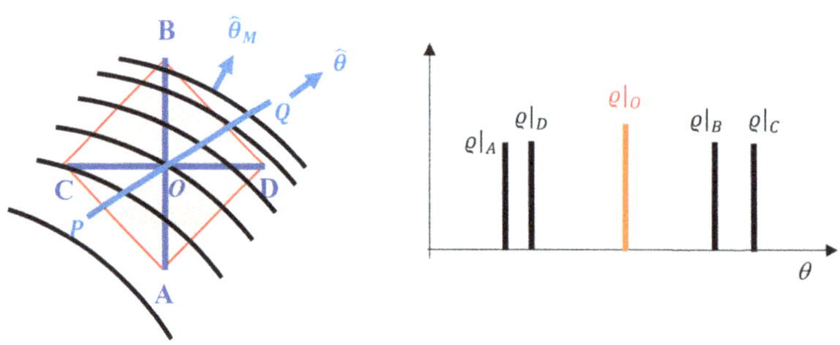

Figure 6.4. If there is no spread in the δ_θ function, equation (6.8) cannot be used to evaluate the density change at the centre O of pixel $ABCD$, since the densities at A, B, C and D would be zero at $\theta = \theta_M|_O$.

where $\hat{\theta} = \cos(\theta_{\vec{AB}} - \theta_v)\hat{s} + \cos(\theta_{\vec{CD}} - \theta_v)\hat{e}$ is the unit vector along the velocity or line-normal direction, $\Delta(\varrho v(\theta))|_{\vec{AB}} = \varrho v(\theta)|_B - \varrho v(\theta)|_A$, and $\Delta(\varrho v(\theta))|_{\vec{CD}} = \varrho v(\theta)|_D - \varrho v(\theta)|_C$.

But with a spread δ_θ function, looking at equation (6.7) again will show that if we were to make the nominal velocity $v(\theta)$ of different θ to be $v \times (\hat{\theta}_M \cdot \hat{\theta})$, then the factor $(\hat{\theta}_M \cdot \hat{\theta})$ would cancel out and then line contents of different θ would travel with the intended velocity v along direction $\hat{\theta}_M$ for the modal line. In practice, we do not need to actually set $v(\theta) \to v \times (\hat{\theta}_M \cdot \hat{\theta})$, but in equation (6.8), just to keep $v(\theta)$ as the intended v for the modal line, and replace $\hat{\theta}$ by $\hat{\theta}_M$, i.e.,

$$\dot{\varrho}^{\text{move}}(\theta)|_{i,j}\,\Delta s \to -[\Delta(\varrho(\theta)v)|_{\overline{AB}}\,\hat{s} + \Delta(\varrho(\theta)v)|_{\overline{CD}}\,\hat{e}] \cdot \hat{\theta}_M \tag{6.9}$$

Using equation (6.9) as the finite-difference solver will make all line contents of different θ in $\varrho(\theta)$ to travel with the same velocity v along direction $\hat{\theta}_M$ for the modal line.

Example 6.1. Prove equation (6.8) from equation (3.16).

Solution: From equation (3.16), by considering $\varrho(\theta)d\theta = \rho(\theta)$, we have, for $d\theta$ at each θ:

$$\dot{\varrho}|_{ABCD}\,\Delta s = -\Delta(\varrho v)|_{\overline{AB}}\cos(\theta_{\overline{AB}} - \theta_v) - \Delta(\varrho v)|_{\overline{CD}}\cos(\theta_{\overline{CD}} - \theta_v) \tag{6.10}$$

where $\Delta(\varrho v)|_{\overline{AB}} = (\varrho v)|_B - (\varrho v)|_A$ and $\Delta(\varrho v)|_{\overline{CD}} = (\varrho v)|_D - (\varrho v)|_C$. With $\hat{\theta}$ defined as the velocity (or line-normal) direction for line orientation θ in figure 3.6,

$$\hat{\theta} = \cos(\theta_{\overline{AB}} - \theta_v)\hat{s} + \cos(\theta_{\overline{CD}} - \theta_v)\hat{e}$$

where $\hat{s}$ is the screw direction (reference direction for θ) and $\hat{e}$ is edge direction, then equation (6.10) can be re-written as:

$$\dot{\varrho}|_{ABCD}\,\Delta s = -[\Delta(\varrho v)|_{\overline{AB}}\,\hat{s} + \Delta(\varrho v)|_{\overline{CD}}\,\hat{e}] \cdot \hat{\theta}$$

which is equation (6.8).

Here,

$$\Delta\vec{J}\,|_{ABCD} = [\Delta(\varrho v)|_{\overline{AB}}\,\hat{s} + \Delta(\varrho v)|_{\overline{CD}}\,\hat{e}] \tag{6.11}$$

may be regarded as an effective flux differential for moving lines across the pixel ABCD, so equation (6.8) may be interpreted as

$$\dot{\varrho}|_{ABCD}\,\Delta s = -\Delta\vec{J}\,|_{ABCD} \cdot \hat{\theta} \tag{6.12}$$

where $\Delta\vec{J}\,|_{ABCD} \cdot \hat{\theta}$ is the projection of the effective flux differential along the $\hat{\theta}$ direction.

6.2 Codes for ADD kinematics simulation

6.2.1 Codes for setting up initial ADD

The following 'discoarse1' and 'cthetacoarse' functions in sequential calls set up the initial ADD of a dislocation line according to equation (6.1) with Gaussian profiles for the δ_r and δ_θ. Q is returned as a 3D matrix $Q(i,j,k)$, where (i,j) mark the meshpoint on the slip plane, and k specifies the position in the θ-space. The 'discoarse1' function also returns a theta(k) array, which is just the nodal points on the θ-space.

```
function [Q, theta] = discoarse1(distance,t_a,sdev,x,line,dtheta,Q)
% This function sets up the delta_r in eqn. (6.1) as Q(i,j,k), for k that
corresponds to the character (theta) of the line at the closest point. Q has
spread as per delta_r, but no spread in k-space.
    h_a=reshape(t_a.',[size(x,2),size(x,2)]);
    distance=reshape(distance.',[size(x,2),size(x,2)]);
% Z is the gaussian profile of the dislocation line over meshpoint space
    Z=1/(sdev*sqrt(2*pi))*exp(-1*((distance).^2)/(2*sdev^2));
    Z(isnan(Z))=0;
% t is cumulative sum of squares of differences of successive points in "line"
    t = cumsum([0;((diff(line.').^2)*ones(2,1)).^(1/4)]).';
% derivative of curve
    derivation=fnder(cscvn(line));
    Q=iter1(Z,t,derivation,Q,h_a);
% setting theta(k)
theta = dtheta*(1:1:size(Q,3));  % theta (dislocation character from XX axis)
increases from ~0 to 2*pi as k increases from 1 to size(Q,3).
end

function [Q]=iter1(Z,t,derivation,Q,h_a)
% This function converts Z into Q(i,j,k) that is a 3D matrix for the density over
spatial (i,j) and theta (k) space.
    for p=1:size(h_a,1)     % here, size(h_a,1) is just size(x)
        for l=1:size(h_a,2)   % here, size(h_a,2) is just size(x); (p,l) are
indices over the meshpoints
        if ~isnan(h_a(p,l))  % if h_a(p,l) is not NaN
         if h_a(p,l)>1
            h_a(p,l)=1;  % h_a is fractional arc length to closest point, thus
this caps any > 1 to 1
         end
            index = find((t/t(end))>=(h_a(p,l)), 1, 'first');
            if index==1
                tangent=[-dot(derivation.coefs((2*index)-
1,:),[0,0,1]),dot(derivation.coefs((2*index),:),[0,0,1])];
            elseif ~isempty(index)
                t0=(h_a(p,l)*t(end))- t(index-1);
                tangent=[-dot(derivation.coefs((2*index)-
3,:),[t0^2,t0^1,1]),dot(derivation.coefs((2*index)-2,:),[t0^2,t0^1,1])];
            end
            theta=2*pi-tan2thet(tangent); % theta (0 to 2*pi) is dislocation line
inclination angle from XX axis
Q(l,p,(floor(theta/(2*pi/size(Q,3))))+1)=Q(l,p,(floor(theta/(2*pi/size(Q,3))))+1)+
Z(p,l); % l,p corrected on Q so i is horizontal axis and j vertical
        end
        end
    end
end

function  QQ = cthetacoarse (Q,dtheta,P)
% This converts Q into QQ that is Gaussian distributed function over theta-space.
QQ=zeros(size(Q,1),size(Q,2),size(Q,3));
for i=1:size(Q,1)
    for j=1:size(Q,2)
        for k=1:size(Q,3)
            if Q(i,j,k)~=0
                for l=round(-.5*size(Q,3)):round(.5*size(Q,3))
                    distance=(abs(l))*dtheta;
                    am=amount(distance,P,dtheta);
```

```
                    if k+l<=0
                      QQ(i,j,k+l+size(Q,3))=QQ(i,j,k+l+size(Q,3))+(Q(i,j,k)*am);
                    elseif k+l>size(Q,3)
                      QQ(i,j,k+l-size(Q,3))= QQ(i,j,k+l-size(Q,3))+(Q(i,j,k)*am);
                    else
                      QQ(i,j,k+l)=QQ(i,j,k+l)+(Q(i,j,k)*am);
                    end
                end
            end
        end
    end
end
end

function [am]=amount(distance,P,dtheta)
% The amount of density as per gaussian function of the difference between θ,
arbitrary character, and θi, dislocation character.
if distance<=5*P
   am=(dtheta/(P*sqrt(2*pi)))*exp(-(distance)^2/(2*(P^2)));
else
   am=0;
end
end
```

6.2.2 Codes for ADD kinematics

The following code will simulate an expanding circle using ADD $\varrho(\vec{r}, \theta)$, in contrast with that given in section 3.4.2 for the approach with $\rho(\vec{r})$ and $\theta(\vec{r})$:

```
%% DDFD_loop: for expanding circle kinematics using ADD
% Preset parameters
:
nthetaintervals=360; % number of nodes in theta-space
dtheta = 2*pi/nthetaintervals;
P=10*dtheta;  % standard deviation for gaussian distribution in the theta space.
:

%% Coarse-graining line into density function with Gaussian profiles in both
spatio and theta spaces
Q=zeros(size(x,2),size(y,2),nthetaintervals); % Density over spatial domain (i,j)
and theta-space (k).
QQ=Q; %% QQ is the working density, and is here initialized to be Q
[xy,distance,t_a]=distance2curve(line.',meshpoints,'linear');
[Q,theta]= discoarse1(distance,t_a,sdev,x,line,dtheta,Q
QQ=cthetacoarse(Q,dtheta,P); % cthetacoarse gives Gaussian spread in theta-space

%% preparing for integration
ds=sqrt(2)*dx;
for i=1:size(x,2)
    for j=1:size(y,2)
        Vbyds(i,j)=vsize/ds;  % setting constant velocity at all spatial points.
    end
end
inbox=ones(size(QQ,1),size(QQ,1));
cos_thetaV(:) = cos(theta(:)+pi/4);  % For rhombus element, see difff2 in 3.4.3.1
sin_thetaV(:) = sin(theta(:)+pi/4);  % CD axis is 45 deg. from XX axis
```

```
%% Integration
for time=1:T/dt
    QQ= difff7(QQ,dt,Vbyds,cos_thetaV,sin_thetaV,inbox);
    %
    % Density QQ can be plotted here, which is skipped for clarity purpose
    %
end


%% difff function to calculate the differentiation over a time-step.
% Executes only when inbox~=0
function [F]= difff7(QQ,dt,Vbyds,cos_thetaV,sin_thetaV,inbox)
G=QQ;
K1=FDM1(G,Vbyds,cos_thetaV,sin_thetaV,inbox);
G=QQ+K1*(dt/2);
K2=FDM1(G,Vbyds,cos_thetaV,sin_thetaV,inbox);
G=QQ+K2*(dt/2);
K3=FDM1(G,Vbyds,cos_thetaV,sin_thetaV,inbox);
G=QQ+K3*dt;
K4=FDM1(G,Vbyds,cos_thetaV,sin_thetaV,inbox);
F=QQ+((K1+(2*K2)+(2*K3)+K4)*dt)/6;
end

% This one uses modal theta for max G as the flux projection angle, eqn. (6.9)
function [K]=FDM1(G,Vbyds,cos_thetaV,sin_thetaV,inbox)
K=zeros(size(G,1),size(G,2),size(G,3));
% modal theta for max G
thetmax=zeros(size(G,1),size(G,2),2);
for i=1:size(G,1)
   for j=1:size(G,2)
       if inbox(i,j)==0


            continue
       end
       if max(G(i,j,:))~=0
            [~,kmax]=max(G(i,j,:));
            thetmax(i,j,1)=cos_thetaV(kmax);
            thetmax(i,j,2)=sin_thetaV(kmax);
       end
   end
end
for i=2:size(G,1)-1
    for j=2:size(G,2)-1
        if inbox(i,j)==0
            continue
        end
        for k=1:size(G,3)
            flux_A=(G(i,j-1,k)*Vbyds(i,j-1)+G(i+1,j-1,k)*Vbyds(i+1,j-
1)+G(i+1,j,k)*Vbyds(i+1,j))/4; % A
            flux_B=(G(i,j+1,k)*Vbyds(i,j+1)+G(i-1,j+1,k)*Vbyds(i-1,j+1)+
G(i-1,j,k)*Vbyds(i-1,j))/4; % B
            flux_C=(G(i-1,j-1,k)*Vbyds(i-1,j-1)+G(i,j-1,k)*Vbyds(i,j-1)+
G(i-1,j,k)*Vbyds(i-1,j))/4; % C
             flux_D=(G(i+1,j+1,k)*Vbyds(i+1,j+1)+G(i,j+1,k)*Vbyds(i,j+1)+
G(i+1,j,k)*Vbyds(i+1,j))/4; % D
            K(i,j,k)=-((flux_B-flux_A)*thetmax(i,j,2)+(flux_D-
flux_C)*thetmax(i,j,1));
        end
    end
end
end
```

Figure 6.5(a) shows the simulation results with the above codes. Here, inside function difff7, the FDM1 function corresponds to equation (6.9), where the flux differential $\Delta\vec{J}\,|_{ABCD}$ is projected along the modal line normal $\hat{\theta}_M$, as a strategy to make all non-modal contents with $\theta \neq \theta_M$ travel with the same speed. The results in figure 6.5(a) indicate that the peak density stays roughly constant as the loop expands, and the total dislocation content is roughly proportional to the loop radius as it should be.

However, if the numerical solver in equation (6.8) is used instead, by dotting $\Delta\vec{J}\,|_{ABCD}$ to the line-normal direction $\hat{\theta}$ of each nodal value θ(k) in the k-space, (see FDM function below), then bizarre results as shown in figure 6.5(b) will be obtained, even though the loop still expands smoothly. Here, the peak density drops significantly as the loop expands, due to the fact that densities of different θ travel at different speeds. The total dislocation content also does not rise in proportional to the loop radius.

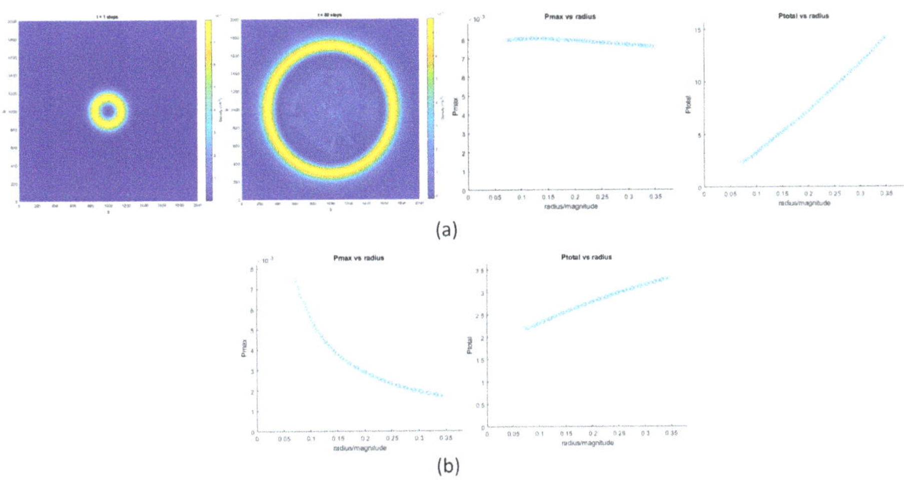

Figure 6.5. (a) Expanding loop simulated using the codes in section 6.2.2, with the FDM1 function in dffff7 that corresponds to equation (6.9). Last two plots show, respectively, the peak density on the loop and the total dislocation content, vs loop radius. (b) Bizarre results obtained with the FDM function that corresponds to equation (6.8), with peak density dropping as the loop expands (left) and total content non-proportional to the loop radius (right).

```
%% This function corresponds to eqn. (6.8). For an expanding loop, it gives
bizarre results in Fig. 6.5(b). This function should not be used for ADD.
function [K]=FDM(G,Vbyds,cos_thetaV,sin_thetaV,inbox)
K=zeros(size(G,1),size(G,2),size(G,3));
for i=2:size(G,1)-1
    for j=2:size(G,2)-1
         if inbox(i,j)==0
             continue
         end
         for k=1:size(G,3)
             flux_A=(G(i,j-1,k)*Vbyds(i,j-1)+G(i+1,j-1,k)*Vbyds(i+1,j-
1)+G(i+1,j,k)*Vbyds(i+1,j))/4; % A
             flux_B=(G(i,j+1,k)*Vbyds(i,j+1)+G(i-1,j+1,k)*Vbyds(i-1,j+1)+G(i-
1,j,k)*Vbyds(i-1,j))/4; % B
             flux_C=(G(i-1,j-1,k)*Vbyds(i-1,j-1)+G(i,j-1,k)*Vbyds(i,j-1)+G(i-
1,j,k)*Vbyds(i-1,j))/4; % C
flux_D=(G(i+1,j+1,k)*Vbyds(i+1,j+1)+G(i,j+1,k)*Vbyds(i,j+1)+G(i+1,j,k)*Vbyds(i+1,j
))/4; % D
             K(i,j,k)=-((flux_B-flux_A)*sin_thetaV(k)+(flux_D-
flux_C)*cos_thetaV(k));
         end
    end
end
end
```

The above example of loop expansion has no rotation, and if rotation is needed, this can be handled by the following functions 'velgrad' and 'thetadiff' called in the integration loop together with difff7:

```
[dVbydxi]=velgrad(QQ,V,rho_min,theta,dx,dy);       % Rotational speed, dV/d(xi),
where xi = dislocation line coordinate for each dislocation character k
QQ= difff7(QQ,dt,V/delta_s,cos_thetaV,sin_thetaV,inbox); % FDM for movement term
[QQ]=thetadiff(QQ,dt,dVbydxi,dtheta);              % FDM for tilting term
```

```
function [dVbydxi]=velgrad(Q,V,rho_min,theta,dx,dy)
nb=10*dx;
dVbydxi=zeros(size(Q,1),size(Q,2),size(Q,3));
xi = zeros(size(theta,2),2);
for k=1:size(Q,3)
    xi(k,1) = cos(theta(k));
    xi(k,2) = sin(theta(k));
    for i=1:size(Q,1)
        for j=1:size(Q,2)
            if Q(i,j,k)<=rho_min
                continue
            end
            dVbydxi(i,j,k)=0;
            n_count=0;
            for ii=i-nb:i+nb
                for jj=j-nb:j+nb
```

```
            if ii>0 && ii<=size(V,1) && jj>0 && jj<=size(V,2) && ii~=i && jj~=j
                        if Q(ii,jj,k)<=rho_min
                            continue
                        end
                        x=(ii-i)*dx; y=(jj-j)*dy;
                        d_xi = x*xi(k,1)+y*xi(k,2);% relative position along disln
                        dot_xi = d_xi/sqrt(x*x+y*y);
                        if abs(abs(dot_xi)-1)>0.01
                            continue
                        end
                        n_count=n_count+1;
                        dVbydxi(i,j,k)=dVbydxi(i,j,k)+(V(ii,jj)-V(i,j))/d_xi;
                    end
                end
            end
            if n_count>0
                dVbydxi(i,j,k)=dVbydxi(i,j,k)/n_count;
            end
        end
    end
end
end

function [Q]=thetadiff(Q,dt,dVbydxi,dtheta)
% R-K Integration
G=Q;
K1=FDM_rotation(G,dVbydxi,dtheta);
G=Q+K1*(dt/2);
G(G<0)=0;
K2=FDM_rotation(G,dVbydxi,dtheta);
G=Q+K2*(dt/2);
G(G<0)=0;
K3=FDM_rotation(G,dVbydxi,dtheta);
G=Q+K3*dt;
G(G<0)=0;
K4=FDM_rotation(G,dVbydxi,dtheta);
Q=Q+((K1+(2*K2)+(2*K3)+K4)*dt)/6;

Q(Q<0)=0;
end

function [K]=FDM_rotation(G,dVbydxi,dtheta)
K=zeros(size(G,1),size(G,2),size(G,3));
pz=size(G,3);
[G_b]=periodic_1d(G, size(G,1), size(G,2), size(G,3), pz);
[dVbydxi_b]=periodic_1d(dVbydxi, size(dVbydxi,1), size(dVbydxi,2),
size(dVbydxi,3), pz);
for i=1:size(G,1)
    for j=1:size(G,2)
        for k=1:size(G,3)
            flux_A=G_b(i,j,k-1+pz)*dVbydxi_b(i,j,k-1+pz);
            flux_B=G_b(i,j,k+1+pz)*dVbydxi_b(i,j,k+1+pz);
            K(i,j,k)=-(flux_B-flux_A)/(2*dtheta);
        end
    end
end
end
```

6.3 Codes for dynamics simulation for ADD

As introduced in section 5.5, two types of interactions are needed in extensive simulation of ADD, namely, (i) short-range Taylor interaction and (ii) long-range elastic interaction. Taylor interaction necessitates a consideration of multiple slip planes or systems, and so before we discuss the interactions, we should first deal with the setting up of the multiple slip systems in the simulation.

6.3.1 MATLAB data structure for multiple slip systems

In terms of data management for simulations on multiple slip systems, the following MATLAB data structure of cell is very useful:

$$\text{parameter}\{\text{slip\_sys}\}$$

where 'parameter' can be a vector (e.g. Burgers vector $\vec{b}$) or a matrix (e.g. density over $\vec{r}$ and θ domains).

Consider a general case of a 3D simulation block as shown in figure 6.6, comprising multiple slip systems. Each slip system (labelled by slip_sys) consists of a stack of parallel slip planes. On each slip plane (labelled by s), there are meshpoints labelled by (i,j). Therefore, each meshpoint in the simulation box in figure 6.6 is labelled by:

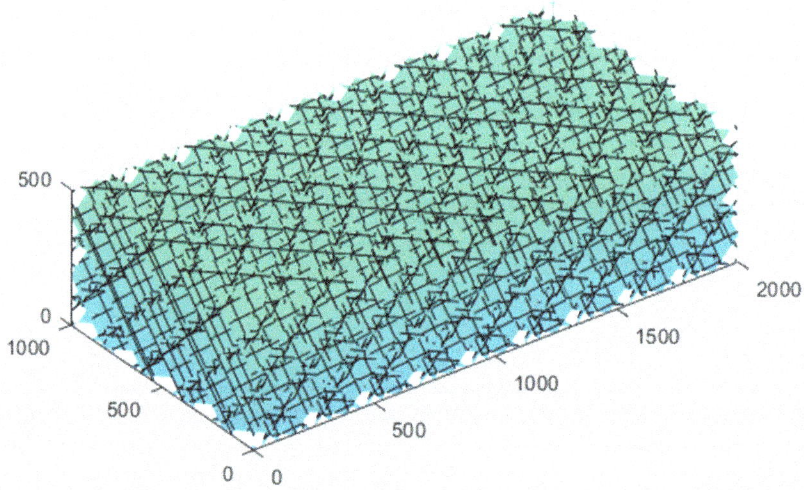

Figure 6.6. A 3D simulation block for FCC with 12 slip systems of ½<110>{1$\bar{1}$1}, each comprising a stack of parallel slip planes. The simulation box has edges along the three cubic axes of the FCC crystal, and dimensions are in units of b = |½<110>|.

$$\{\text{slip\_sys}\}(s, i, j) \tag{6.13}$$

As can be seen from figure 6.6, the slip planes {slip_sys}(s) have different sizes inside the simulation box, because they are flat planes truncated by the simulation box. During the simulation, dislocation density will be made to evolve on each slip plane as per strategies discussed so far, hence it will be useful to leave the slip planes larger than the simulation box, while we label which meshpoints are inside the simulation box. Therefore, using the cell data structure above, attributes of the meshpoints will include the following:

XG{slip_sys}(s, i, j)
: x–coordinate of meshpoint in global box
YG{slip_sys}(s, i, j)
: y–coordinate of meshpoint in global box
ZG{slip_sys}(s, i, j)
: z–coordinate of meshpoint in global box
Inbox_flag{slip_sys}(s, i, j)
: equals 1 if meshpoint is inside global box, 0 otherwise (6.14)

For some operations in the main programme (e.g., see section 6.3.2), it will be more convenient to just loop over the meshpoints inside the simulation box without caring about those falling outside. In this representation, the meshpoints inside the simulation box are labelled by the index site_indx, which runs from 1 to n_site, the latter being the number of meshpoints inside the simulation box. Each simulation-box meshpoint will have the following attributes:

ite{site_indx}
: the[s, i, j, slip_sys]for the meshpoint specified by site_indx
G(1: 3, site_indx)
: the global–box coordinates(1: 3)of the meshpoint specified by site_indx (6.15)

Finally, other operations may just require looping over the simulation-box meshpoints on each slip plane. Therefore, for the slip plane specified by {s,slip_sys}, the simulation-box meshpoints are sequentially labelled by an index site_on_sp_indx, and the following cell records the site_indx indices of the meshpoints in the global box numbering system:

$$\text{site\_indx\_on\_sp}\{s, \text{slip\_sys}\}(\text{site\_on\_sp\_indx}) = \text{site\_indx} \tag{6.16}$$

The following 'globalmaker' function sets up the simulation block in figure 6.6, by returning the meshpoint attributes described above.

```
% Declaration of relevant parameters in main programme
magnitudex=2000*b;  % size of global simulation cell in x direction
magnitudey=1000*b;  % size of global simulation cell in y direction
magnitudez=500*b;   % size of global simulation cell in z direction
globalcell = [1 0 0; 0 1 0; 0 0 1];     % directions of global cell
nxintervals = 41;   % number of x intervals
nyintervals = 41;   % number of y intervals

:

% Get specifics of the slip systems in the crystal structure studied
[n_slip_sys, n_vec, s_vec, e_vec, b_vec, tau_app]= slipsystem('fcc', b,
stress_app_state);

:

% Call globalmaker function
[XG,YG,ZG,inbox_flag,n_site,site,rG,site_indx_on_sp]=globemaker(seq,n_slip_sys,mag
nitudex,magnitudey,magnitudez,nxintervals,nyintervals,dx,dy,ds,n_vec,s_vec,e_vec,g
lobalcell);

:

function[XG,YG,ZG,inbox_flag,n_site,site,rG,site_indx_on_sp]=
globemaker(seq,n_slip_sys,magnitudex,magnitudey,magnitudez,nxintervals,nyintervals
,dx,dy,ds,n_vec,s_vec,e_vec,globalcell)

XG=cell(n_slip_sys);  % (XG,YG,ZG){slip_sys}(s,i,j) = global-box coordinates
YG=cell(n_slip_sys);  % of meshpoint(i,j) on plane s in slip system slip_sys
ZG=cell(n_slip_sys);
inbox_flag=cell(n_slip_sys); % inbox_flag{slip_sys}(s,i,j)=1 if

for slip_sys=1:n_slip_sys
    X=zeros(seq,nxintervals,nyintervals);
    Y=zeros(seq,nxintervals,nyintervals);
    Z=zeros(seq,nxintervals,nyintervals);
    for s=1:seq
        for i=1:nxintervals
            for j=1:nyintervals
                p=n_vec{slip_sys}*ds*(s-seq/2) + (i-
nxintervals/2)*dx*s_vec{slip_sys} + (j-nyintervals/2)*dy*e_vec{slip_sys};
% p is vector of meshpoint(i,j) on plane s relative to centre of plane
                X(s,i,j)=dot(p(:),globalcell(1,:))+magnitudex/2;
                Y(s,i,j)=dot(p(:),globalcell(2,:))+magnitudey/2;
                Z(s,i,j)=dot(p(:),globalcell(3,:))+magnitudez/2;
% (X,Y,Z)=global-box coordinates of meshpoint (i,j) on plane s
            end
        end
    end
    XG{slip_sys}=X; YG{slip_sys}=Y; ZG{slip_sys}=Z;

    inbox=ones(seq,nxintervals,nyintervals); % flag to indicate region of each
slip plane inside global simulation box
    inbox(X>magnitudex | X<0)=0; % set flag to 0 for nodes
    inbox(Y>magnitudey | Y<0)=0; % outside global simulation box
    inbox(Z>magnitudez | Z<0)=0;
    inbox_flag{slip_sys}=inbox;
end
```

```
% Label sites in global box with inbox_flag = 1, get site info and coordinates
site_indx_on_sp=cell(seq,n_slip_sys);
site_indx = 0;
for slip_sys = 1:n_slip_sys
    for s = 1:seq
        site_on_sp_indx = 0;
        for i = 1:nxintervals
            for j = 1:nyintervals
                if inbox_flag{slip_sys}(s,i,j) == 1
                    site_indx = site_indx + 1;
                    site_on_sp_indx=site_on_sp_indx + 1;
                    site{site_indx}=[s, i, j, slip_sys];
                    rG(1,site_indx)=XG{slip_sys}(s,i,j); % Site positions
                    rG(2,site_indx)=YG{slip_sys}(s,i,j); % in global index
                    rG(3,site_indx)=ZG{slip_sys}(s,i,j);
                    site_indx_on_sp{s,slip_sys}(site_on_sp_indx)=site_indx;  %
Collate site index on each slip plane for difff
                end
            end
        end
    end
end
n_site = site_indx;     % number of sites within global box
end
```

6.3.2 Code for setting up the initial ADD

Using the cell data structure introduced in section 6.3.1, we may represent the ADD $\varrho(\vec{r}, \theta)$ defined on both the $\vec{r}$- and θ-space, as follows:

$$\text{density\{slip\_sys\}(s, i, j, k)} \tag{6.17}$$

where
slip_sys specifies the slip system,
s specifies the slip plane,
(i,j) specify the meshpoint in the $\vec{r}$-space, and
k specifies the node in the θ-space

The following function sets up the initial ADD on each of a number (seq) of slip planes. On each slip plane {slip_sys}(s), a number (nodisp) of dislocation loops with random radius are set up according to Gaussian profiles in equation (6.3). The resultant Q(s,i,j,k) is then loaded to density{slip_sys}(s,i,j,k).

```
% In main programme
:
for slip_sys=1:n_slip_sys
    [Q]=
densityfinder1(nodisp,seq,magnitude,magnitudey,Q,meshpoints,sdev,dtheta,P,x,rho_sc
ale);
    density{slip_sys}=Q;
end

:

function [Q]=
densityfinder1(nodisp,seq,magnitude,magnitudey,Q,meshpoints,sdev,dtheta,P,x,rho_sc
ale)
% Q(s,i,j,k) has four indices; s specifies the slip plane, (i,j) specifies
% meshpoint on the slip plane, k specifies disln character (theta) space.
% nodisp = number of dislocations on each slip plane
% seq = number of slip planes
% magnitude = size of the 2D plane

for s=1:seq
    R=zeros(size(Q,2),size(Q,3),size(Q,4)); % local density on the slip plane.
    for i=1:nodisp
        xCenter = 0.5*magnitude;
        yCenter = 0.5*magnitude;
        xRadius = 0.3*magnitudey + 0.25*rand()*magnitude; % random radius
        yRadius =xRadius;
        theta = 0:0.01:2*pi;
        lx=xRadius*cos(theta) + xCenter;
        ly=yRadius*sin(theta)+ yCenter;
        lines{i}=[lx;ly];
        [~,distance,t_a]=distance2curve(lines{i}.',meshpoints ,'linear');
        [R,~]=discoarse1(distance,t_a,sdev,x,lines{i},dtheta,R);
        % RR is the density matrix on each slip plane as a local matrix
        RR=cthetacoarse(R,dtheta,P); % distributes density over k space according
to gaussian distribution for delta-theta function.
    end
    Q(s,:,:,:)=RR*rho_scale; % scale density to realistic magnitude
end
end
```

6.3.3 Codes for Taylor interactions

Calculating the Taylor interactions according to equation (5.23) requires finding the interacting neighbours of each meshpoint in the system. In the ‘interaction_lr_core1’ function for evaluating the Mura interaction coefficients discussed in section 4.4.2, the interacting neighbours (is,js) within cutoff radii nx_co and ny_co from the current meshpoint (i,j) are easily called out in the following manner because the meshgrid is rectangular:

```
for i=1:size(rhoG,1)
    for j=1:size(rhoG,2)
        for ii=1:nx_co*2+1      % local index in neighbourhood of i
            is=px+i+(ii-nx_co-1);     % index relative to global origin
            for jj=1:ny_co*2+1
                js=py+j+(jj-ny_co-1);
                    % calculate interacting coefficients here

            end

        end
    end
end
```

This method, unfortunately, no longer works for the simulation cell in figure 6.6 with multiple slip systems in 3D, because the slip systems are oblique to each another so that the neighbours on different slip systems of a given meshpoint {slip_sys}(s,i,j) (see equation (6.13)) cannot be called out in the above manner. The fix is the method of neighbour list, which is frequently employed in MD simulations (Allen and Tildesley 2017). In this method, the neighbour relationships of the meshpoints are found once at the beginning, after 'globalmaker' has set up the simulation box, and are recorded for all subsequent call outs. For Taylor interactions that are short ranged with a small neighbourhood cutoff Rc on the order of the pixel size, the neighbour list can be built rather quickly by the following function 'neighbourlist_Taylor', even though the simulation cell has a large n_site in tens of thousands. The function returns the neighbour list in the following format:

$$\text{nbr\_list\_Taylor\{site\_indx\}\{nb\}:thesiteindexofthenb}^{\text{th}}\text{neighbouroftheglobal–box meshpointofsiteindex“site\_indx”withincutoffRc} \tag{6.18}$$

```
function [nbr_list_Taylor]=neighbourlist_Taylor(n_site,rG,Rc)
% Build neighbour list of each site for Taylor hardening interaction
% For each site, find each of its neighbour within Rc and store its site info.
for site_indx=1:n_site
    nb=0;        % neighbour number
    for site_indx_nb=1:n_site
        if site_indx_nb~=site_indx
            Rvec(:)=rG(:,site_indx_nb)-rG(:,site_indx);  % R = r'-r
            if sqrt(dot(Rvec,Rvec))<=Rc
                nb=nb+1;
                nbr_list_Taylor{site_indx}{nb}=site_indx_nb;
            end
        end
    end
end
end
```

After finding the neighbour list, the Taylor interactions may be calculated by the following function 'interaction_Taylor', which is called in the integration loop as shown below:

```
[tau_int_Taylor]=interaction_Taylor(density,nbr_list_Taylor,site,rhosum,slip_sys,s
,site_indx_on_sp,Taylor_coeff,mu,b,n_vec);


function [tau_int_Taylor]=interaction_Taylor(density, nbr_list_Taylor, site,
rhosum, slip_sys, s, site_indx_on_sp, Taylor_coeff, mu, b, n_vec)

tau_int_Taylor=zeros(size(density{slip_sys},2),size(density{slip_sys},3));

% Establishing short-ranged Taylor hardening interaction RSS for each field point
% (i,j) on slip plane s

for i=1:size(site_indx_on_sp{s,slip_sys},2)
    site_indx=site_indx_on_sp{s,slip_sys}(i);
    rho_eff=0;                          % effective density, tau_Taylor =
mu*b*sqrt(rho_eff)
    for nb=1:size(nbr_list_Taylor{site_indx},2)
        site_indx_nb=nbr_list_Taylor{site_indx}{nb};
        slip_sys_nb=site{site_indx_nb}(4);
        if n_vec{slip_sys}==n_vec{slip_sys_nb}
            rho_eff=rho_eff+Taylor_coeff(1)*rhosum{site_indx_nb};
        else
            rho_eff=rho_eff+Taylor_coeff(2)*rhosum{site_indx_nb};
        end
    end
    tau_int_Taylor(site{site_indx}(2),site{site_indx}(3))=mu*b*sqrt(rho_eff);
end
end
```

Taylor interactions involve the total dislocation density $\rho_{\text{total}}^{\alpha'}$ or 'rhosum' (see equation (5.23)), which may be calculated as follows:

```
function [rhosum]=find_rhosum(density,site,n_site)
% find the sum of density over theta space at each site within global box
rhosum=cell(n_site);
for site_indx=1:n_site
    s = site{site_indx}(1);
    i = site{site_indx}(2);
    j = site{site_indx}(3);
    slip_sys = site{site_indx}(4);
    rhosum{site_indx} = sum(density{slip_sys}(s,i,j,:));
end
```

6.3.4 Codes for Mura interactions

Just like for Taylor interactions, for Mura (long-range elastic) interactions, the corresponding neighbour list needs to be constructed before the integration loop in the main programme. For efficient calculation of the Mura interaction coefficients $g_{\alpha\alpha' h}(\vec{r} - \vec{r}')$ in equation (4.21B), more information of the neighbours than simply their site indices is required. Since the relative vector $\vec{R} = \vec{r} - \vec{r}'$ between the field and source points is involved in the $g_{\alpha\alpha' h}$ coefficients, but the neighbourhoods of the meshpoints should be identical despite not being a simple rectangular lattice, it will be advantageous to build a common list (Rvec_list) of the $\vec{R}$ vectors of the neighbours that every field meshpoint $\vec{r}$ will see, so that the $g_{\alpha\alpha' h}$'s can be computed more efficiently using equation (4.21B). Then, for the nb^{th} neighbour of a given meshpoint of site index 'site_indx' in the global box, we just need to record the index (Rvec_indx) of the $\vec{R}$ vector in the common neighbourhood list, and then the $\vec{R}$ can be called out easily when computing the $g_{\alpha\alpha'h}$'s. Such information is generated by the function 'neighbourlist_elastic_int' below, which returns the following:

Rvec_list{i}: the ith $\vec{R}=\vec{r}-\vec{r}'$ vector in the common neighbourhood of the meshgrid environment of the simulation box with in cut off Rc_ int ;

n_Rvec: total number of $\vec{R}$ vectors in the above common neighbourhood (i.e. length of the Rvec_list); (6.19)

nbr_list{site_indx}{nb} = [site_indx_nb, Rvec_indx] returns two parameters: site_indx_nbis the site index of the nb^{th}neighbour of the global–box meshpoint of site index"site_indx", and Rvec_indx is the index of that neighbour in the common neighbourhood list Rvec_list.

```
function [nbr_list,Rvec_list,n_Rvec]=neighbourlist_elastic_int(n_site,rG,Rc_int)
% Build neighbour list of each site
% For each site, find each of its neighbour within Rc_int and (1) store its site
index and neighbour distance,
% (2) establish a list of neighbour relative positions and label the
% neighbour according to the relative position in the Rvec_list (for
% g_alpha,alpha')
Rvec=zeros(1,3);   % relative position of neighbour, R = r' - r
n_Rvec=0;   % number of non-repeating relative positions of neighbours
Rvec_list{1}=[0 0 0]; % list of non-repeating relative positions of neighbours
for site_indx=1:n_site
    nb=0;   % neighbour number
    for site_indx_nb=1:n_site
        if site_indx_nb~=site_indx
            Rvec(:)=rG(:,site_indx_nb)-rG(:,site_indx);  % R = r'-r
            if sqrt(dot(Rvec,Rvec))<=Rc_int
                nb=nb+1;
[Rvec_list,Rvec_indx,n_Rvec]=neighbour_position_list(Rvec_list,n_Rvec,Rvec); %
check whether Rvec is already in Rvec_list, if not, add it there
                nbr_list{site_indx}{nb}=[site_indx_nb, Rvec_indx];
            end
        end
    end
end
end

function
[Rvec_list,Rvec_indx,n_Rvec]=neighbour_position_list(Rvec_list,n_Rvec,Rvec)
```

```
% check whether input Rvec is already in Rvec_list, if not, add it there
found = 0;
if n_Rvec==0
    n_Rvec=n_Rvec+1;
    Rvec_list{n_Rvec}=Rvec;
    Rvec_indx=n_Rvec;
else
    for i=1:n_Rvec
        r=Rvec-Rvec_list{i};
        if dot(r,r)<1e-4
            Rvec_indx=i;
            found = 1;
            break
        end
    end
    if found == 0
        n_Rvec=n_Rvec+1;
        Rvec_list{n_Rvec}=Rvec;
        Rvec_indx=n_Rvec;
    end
end
end
```

After the neighbour list is built, the Mura interaction coefficients $g_{\alpha\alpha' h}$'s can be computed using the following function 'interaction_coeff3D':

```
function [int_c_lr3D]=interaction_coeff3D(n_vec, b_vec, s_vec, n_slip_sys, a, dx,
mu, nu, Rvec_list, n_Rvec)
% Long-range interaction coefficients in Mura's formulation for interaction
% stress in 3D for multi-slip
% int_c_lr3D{i_Rvec}(h) = n_p(2)*s_q(2)*C_pqrs/mu*C_klij*b_i(1)*e_sjh*G_rk,l for
% each Rvec vector specified by i_Rvec in 3D space within cutoff distance,
% with h specifying the source-dislocation line direction (i.e. d_l(h)).
% a is Cai et al.'s atomic size that removes singularity in Mura's formula
% Output stored as cell array:
int_c_lr3D{i_Rev,slip_sys_source,slip_sys_field}(1:3)=a
% vector of three
int_c_lr3D=cell(n_Rvec,n_slip_sys,n_slip_sys);
lambda=2*mu*nu/(1-2*nu);
lom=lambda/mu;
[e]=epsi; % permutation tensor
% Elastic constants
% C_ijkl = mu*(delta_ik * delta_jl + delta_il * delta_kj) + lambda*(delta_ij *
delta_kl)
cbymu = zeros(3,3,3,3);
for i = 1:3
    for j = 1:3
        for k = 1:3
            for l = 1:3
                cbymu(i,j,k,l) = (i==k)*(j==l)+(i==l)*(k==j)+lom*(i==j)*(k==l);
            end
        end
    end
end
```

```
%Long range interaction coefficients
for slip_sys2=1:n_slip_sys    % the field one
    n2=n_vec{slip_sys2}; b2=s_vec{slip_sys2};
    for slip_sys=1:n_slip_sys
        b=b_vec{slip_sys};
        for Rvec_indx=1:n_Rvec
            x=Rvec_list{Rvec_indx};  % x(3) is the Rvec=(r'-r) vector in space
within cutoff,
% int_c_lr3D{i_Rvec,slip_sys,slip_sys_field}(h) =
n_p(2)*s_q(2)*C_pqrs/mu*C_klij*b_i(1)*e_sjh*G_rk,l
% for each Rvec within cutoff specified by i_Rvec, n_vec and b_vec specified by
slip_sys
            R=sqrt(dot(x,x));
            R_revo=(R^2+(a*dx)^2)^0.5;   % Cai et al.'s singularity removal

            g=zeros(1,3);
            for j=1:3
                for s=1:3
                    for r=1:3
                        for p=1:3
                            for q=1:2
                                % t1 = n_p(2) * b_q(2) * C_pqrs / mu;
                                t1=n2(p)*b2(q)*cbymu(p,q,r,s);

                                if t1~=0
                                    for i=1:2
                                        for k=1:3
                                            for l=1:3
                                                % t2 = b_i * C_ijkl * G_rk,l


                                                t2=-1/(16*pi*(1-
nu))*b(i)*cbymu(i,j,k,l)* ...
                                                    ((r==k)*3*(a*dx)^2*2*(1-
nu)/R_revo^5*x(l)+... % extra term from Cai et al.
                                                    (3-
4*nu)*(r==k)/R_revo^3*x(l)+3*x(r)*x(k)*x(l)/R_revo^5- ...

((r==l)*x(k)+(k==l)*x(r))/R_revo^3);

                                                for h=1:3
                                                    g(h)=g(h)+e(s,j,h)*t2*t1;
                                                end
                                            end
                                        end
                                    end
                                end
                            end
                        end
                    end
                end
            end
            int_c_lr3D{Rvec_indx,slip_sys,slip_sys2}=g;
        end
    end
end
end
```

Then, the long-range elastic interaction may be computed using function ‘interaction_lr3D’, which is called in the integration loop as:

```
[tau_int_lr]=interaction_lr3D(density,nbr_list,site,GND_vec,slip_sys,s,int_c_lr3D,
dV,site_indx_on_sp);
function [tau_int_lr]=interaction_lr3D(density, nbr_list, site, GND_vec, slip_sys,
s, int_c_lr3D, dV, site_indx_on_sp)

tau_int_lr=zeros(size(density{slip_sys},2),size(density{slip_sys},3));

% Establishing long-range interaction RSS for each field point
% (i,j) on slip plane s

for i=1:size(site_indx_on_sp{s,slip_sys},2)
    site_indx=site_indx_on_sp{s,slip_sys}(i);
    t=0;
    for nb=1:size(nbr_list{site_indx},2)
        site_indx_nb=nbr_list{site_indx}{nb}(1);
        Rvec_indx=nbr_list{site_indx}{nb}(2);
        slip_sys_nb=site{site_indx_nb}(4);
        g=int_c_lr3D{Rvec_indx,slip_sys_nb,slip_sys};
        gnd=GND_vec{site_indx_nb};
        t=t+dot(g,gnd);
    end
    tau_int_lr(site{site_indx}(2),site{site_indx}(3))=t;
end
tau_int_lr(:,:)=tau_int_lr(:,:)*dV;
end
```

The above involves the GND (see equations (5.24) and (5.25)), which is calculated as follows:

```
function [GND_vec]=find_gnd(density,site,n_site,theta,s_vec, e_vec)
% find the GND at each site within global box
GND_vec=cell(n_site);
for site_indx=1:n_site
    s = site{site_indx}(1);
    i = site{site_indx}(2);
    j = site{site_indx}(3);
    slip_sys = site{site_indx}(4);
    gx = 0;
    gy = 0;
    for k = 1:size(density{slip_sys},4)
        gx = gx + density{slip_sys}(s,i,j,k)*cos(theta(k));   % i (horizontal)
axis is the screw axis (reference for dislocation character)
        gy = gy + density{slip_sys}(s,i,j,k)*sin(theta(k));
    end
    GND_vec{site_indx}=gx*s_vec{slip_sys}+gy*e_vec{slip_sys};
end
```

6.3.5 Codes for dynamics simulation of 3D multi-slip

Finally, the main programme that executes all of the above is given below:

```
%% DDFD_extensive_multislip
% Material properties
mu=1;                    % Shear modulus
nu=0.3;                  % Poisson's ratio
Taylor_coeff=[0.1 0.3]; % Taylor hardening coeffs [alpha|| alpha|_] (first for
same slip plane, second for different slip plane)
mobility = 5000;         % dislocation mobility in units of b/dt: v=mobility*tau/mu

% Global simulation cell
b = 1;                   % Burgers vector mangitude, 1/sqrt(2)
magnitudex=2000*b;       % size of global simulation cell in x direction
magnitudey=1000*b;       % size of global simulation cell in y direction
magnitudez=500*b;        % size of global simulation cell in z direction
globalcell = [1 0 0; 0 1 0; 0 0 1];  % directions of global cell

% Slip plane simulation domain
magnitude=magnitudex*1.2;  % size of global simulation cell in x direction
nxintervals = 41;          % number of x intervals
nyintervals = 41;          % number of y intervals
ds = magnitudex/16;        % slip plane thickness
seq=ceil(max(magnitudex*sqrt(3)/ds,magnitudez*sqrt(3)/ds)); % # of parallel planes

% Applied quantities
dt=1;                    % time step
T=200*dt;                % time to finish
strain_exit = 0.1;       % exit magnitude of strain(1,1)
stress_peierls = 0.0002;   % Peierls stress in units of mu
stress_rate = -(0.002+stress_peierls)/T;  % applied stress rate
stress_app_state = [1 0 0; 0 0 0; 0 0 0]; % appl. stress tensor in crystal system
Rc_int=30*b;             % cutoff radius of dislocation interactions
rho_min=1e-2;            % minimum density to skip long-range interaction
a = 1;                   % Cai et al.'s singularity removal length scale

% Simulation parameters
thetas = 0;
thetaf = pi/2;
nthetaintervals=360;     % number of nodes over theta-space
dtheta = 2*pi/nthetaintervals;  % resolution of theta space for density storage
P=10*dtheta;  % standard deviation for gaussian distribution in theta space
sdev=10*b;    % standard deviation for the gaussian profile in spatial domain
rho_scale=1e-4;  % factor to scale the density generated by densityfinder
rho_max=rho_scale*100;  % max density allowed to prevent instability

% Get slip system for FCC
[n_slip_sys, n_vec, s_vec, e_vec, b_vec, tau_app]= slipsystem('fcc', b,
stress_app_state);

%% Common mesh grid for slip systems (horizontal axis is i, vertical is j)
x=linspace(0*magnitude,1*magnitude,nxintervals);
y=linspace(0*magnitude,1*magnitude,nyintervals);
dx=(x(end)-x(1))/(size(x,2)-1);
dy=(y(end)-y(1))/(size(y,2)-1);
meshpoints=zeros((size(x,2))*size(y,2),2);
k = 0;
for i=1:size(x,2)
    for j=1:size(y,2)
        k = k+1;
        meshpoints(k,1)=x(1,i);
        meshpoints(k,2)=y(1,j);
```

```
    end
end

%% This block is for establishing the dislocation lines in slip-system meshgrid
nodisp=5;   % number of dislocations on each slip plane
Q=zeros(seq,size(x,2),size(y,2),nthetaintervals); % ADD
density=cell(n_slip_sys);   % density is a 1:12 cell array

for slip_sys=1:n_slip_sys
    [Q]=
densityfinder1(nodisp,seq,magnitude,magnitudey,Q,meshpoints,sdev,dtheta,P,x,rho_sc
ale);
    density{slip_sys}=Q;
end

%% Global simulation cell mesh grid
[XG,YG,ZG,inbox_flag,n_site,site,rG,site_indx_on_sp]=globemaker(seq,n_slip_sys,mag
nitudex,magnitudey,magnitudez,nxintervals,nyintervals,dx,dy,ds,n_vec,s_vec,e_vec,g
lobalcell);

%% preparing for integration
% Build neighbour list of each site for elastic interactions
[nbr_list,Rvec_list,n_Rvec]=neighbourlist_elastic_int(n_site,rG,Rc_int);
% Elastic interaction coefficient
% g(alpha',alpha)=int_c_lr3D{i_Rvec_source,slip_sys_source,slip_sys_field}(1:3)
[int_c_lr3D]=interaction_coeff3D(n_vec, b_vec, s_vec, n_slip_sys, a, dx, mu, nu,
Rvec_list, n_Rvec);
dV=dx*dy*ds;

% Build neighbour list of each site for Taylor hardening interactions
[nbr_list_Taylor]=neighbourlist_Taylor(n_site,rG,max(dx,dy));

% setting theta(k)
theta = dtheta*(1:1:nthetaintervals);
cos_thetaV(:) = cos(theta(:)+pi/4);
sin_thetaV(:) = sin(theta(:)+pi/4);  % CD axis is 45 deg. from XX axis

%% PDE solving
tau_int_lr=zeros(size(x,2),size(y,2));
tau_int_Taylor=zeros(size(x,2),size(y,2));
tau_eff=zeros(size(x,2),size(y,2));
QQ=zeros(size(x,2),size(x,2),nthetaintervals);
strain=zeros(3,3);
delta_s=sqrt(2)*dx;
for time=1:T/dt
    stress=stress_rate*time*dt;
    [GND_vec]=find_gnd(density,site,n_site,theta,s_vec,e_vec);
    [rhosum]=find_rhosum(density,site,n_site);
    [density]=free_boundary(n_slip_sys,density,inbox_flag);
    density0=density;
    for slip_sys = 1:n_slip_sys
        RSS=tau_app{slip_sys}*stress;
        inbox_sys=inbox_flag{slip_sys};
        [b_diad_n]=diad(b_vec{slip_sys},n_vec{slip_sys});
        for s=1:seq
            inbox(:,:)=inbox_sys(s,:,:);
            QQ(:,:,:)=density0{slip_sys}(s,:,:,:);
[tau_int_lr]=interaction_lr3D(density,nbr_list,site,GND_vec,slip_sys,s,int_c_lr3D,
dV,site_indx_on_sp);
```

```
[tau_int_Taylor]=interaction_Taylor(density,nbr_list_Taylor,site,rhosum,slip_sys,s
,site_indx_on_sp,Taylor_coeff,mu,b,n_vec);
tau_eff(:,:)=abs(RSS+tau_int_lr(:,:))-(stress_peierls+tau_int_Taylor(:,:));
            tau_eff(tau_eff<0)=0;
            tau_eff(:,:)=sign(RSS+tau_int_lr(:,:))*tau_eff(:,:);
            V(:,:)=tau_eff(:,:)*mobility;  % Mobility assumed independent of theta
            [dVbydxi]=velgrad(QQ,V,rho_min,theta,dx,dy);  % Rotational speed
            QQ= difff7(QQ,dt,V/delta_s,cos_thetaV,sin_thetaV,inbox); % movement
            [QQ]=thetadiff(QQ,dt,dVbydxi,dtheta);    % tilting
            QQ(QQ<0)=0;
            QQ(QQ>rho_max)=rho_max;  % cap density to avoid instability
            density{slip_sys}(s,:,:,:)=QQ(:,:,:);
[strain]=strain_finder(strain,QQ,V,slip_sys,dt,b_diad_n,s_vec,e_vec,s,site_indx_on
_sp,site,n_site,theta);
        end
    end
    if abs(strain(1,1)) > strain_exit  % check for exit
        break
    end
end
```

Figure 6.7 shows the simulation result from running the above code, for the simulation model shown in figure 6.6. Here, the face-centred cubic (FCC) crystal has 12 slip systems of ½<110>{1$\bar{1}$1}, and the simulation box measures 2000b × 1000b × 500b, where b is the magnitude of the ½<110> Burgers vector, with edges along the three <100> cubic axes of the FCC crystal. Each parallel stack of coarse-grained planes has interplanar spacing of 125b, and on each slip plane, three ½<110> slip vectors are present according to the ½<110>{1$\bar{1}$1} slip systems. The number of meshpoints in the simulation box, n_site, is 26 904 which is manageable by a notebook computer. The six faces of the box are free boundaries where dislocations

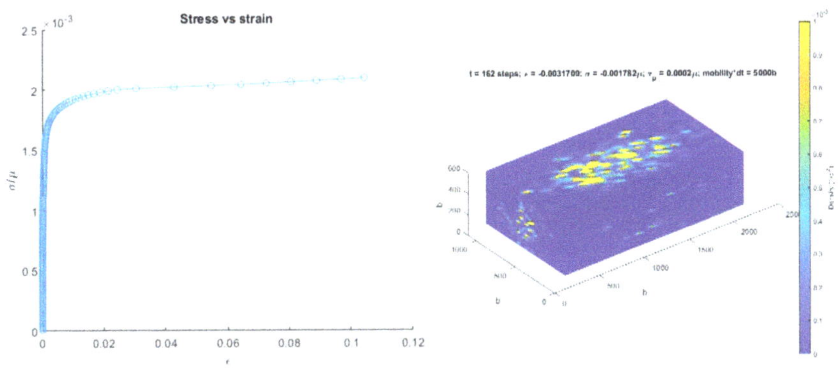

(a) (b)

Figure 6.7. Multi-slip extensive simulation in 3D for FCC with 12 slip systems of ½<110>{1$\bar{1}$1}, with the coarse-grained slip planes shown in figure 6.6. (a) Simulated plastic stress–strain curve under compressive stress along the [100] cubic axis of the FCC crystal at rate $1.1\times10^{-5}\,\mu$ per time step, with free boundary conditions on the six faces of the simulation box. Panel (b) shows snapshot of the dislocation density summed over the θ space on the three free surfaces of the simulation box at around 0.3% strain. (Reproduced from Kalaei *et al* (2022), copyright (2022) with permission from Elsevier.)

can escape. Other parameters are set in the codes above. Figure 6.7(a) shows the simulated stress–strain curve for the direct stress and strain along the stress application direction of [100]. Figure 6.7(b) shows that, as plastic strains develop, dislocation entanglements occur due to the Taylor and Mura interactions leading to high local densities. Work-hardening in figure 6.7(a) is weak because there is no dislocation source placed in the simulation, and only the initially set dislocations evolve. Therefore, once the dislocations begin to move, the strain rapidly builds up without much resistance from dislocation multiplication. How to build dislocation source that is compatible with the kinematics law in equation (5.12) will be an interesting future problem to explore. Note that in the results in figure 6.7, elastic deformation has not been added.

References

Allen M P and Tildesley D J 2017 *Computer Simulation of Liquids* 2nd edn (Oxford: Oxford University Press)

Kalaei A, Xiang Y and Ngan A H W 2022 An efficient and minimalist scheme for continuous dislocation dynamics *Int. J. Plast.* **158** 103433

IOP Publishing

Continuous Dislocation Dynamics Made Simple

Alfonso Ngan and Alireza Kalaei

Chapter 7

Concluding remarks

7.1 Earlier CDD models revisited

To let the reader understand CDD without unnecessary delay, the literature review presented in chapter 1 was rather brief. Here, we present a more comprehensive review of previous CDD models, so as to highlight the importance of the ADD.

7.1.1 Exact CDD for intensive dislocations as GNDs

As discussed in chapter 3, the exact evolution law for GNDs as intensive (i.e. single-dislocation) microstructures exists in several forms, namely,

- For a Nye tensor

$$\text{Maxwell law: } \dot{\boldsymbol{\alpha}} = -\vec{\nabla} \times (\boldsymbol{\alpha} \times \vec{v}) \tag{3.5}$$

- For intensive dislocations as GNDs:

$$\text{Maxwell form:} \quad \dot{\vec{\rho}} = -\vec{\nabla} \times (\vec{\rho} \times \vec{v}) \tag{3.4}$$

$$\text{Movement} - \text{tilting form:} \quad \dot{\vec{\rho}} = -[\vec{\nabla}(\rho v) \cdot \hat{v}] \quad \hat{\xi} + \rho(\nabla_{\hat{\xi}} v) \quad \hat{v} \tag{3.9}$$

$$\text{Divergence} - \text{curvature form:} \quad \dot{\rho} = -\vec{\nabla} \cdot (\rho \vec{v}) + \rho v k \tag{3.11}$$

$$\text{Finite} - \text{difference form:} \quad \dot{\rho} \Delta s \sin(\theta_s - \theta) = -\Delta(\rho v)|_{\Delta \vec{s}} \tag{3.13}$$

$$\dot{\theta} = \nabla_{\hat{\xi}} \ v \tag{3.9B}$$

So, naturally, there have been a number of CDD models developed for GNDs as intensive dislocations. For instance, Varadhan *et al* (2006) solved the Maxwell law in equation (3.5) using a Galerkin-least square method to model intensive dislocation structures such as the Frank-Read source, but they set

doi:10.1088/978-0-7503-6330-3ch7

the dislocation velocity to be constant instead of considering full dynamics with mutual long-range interactions. Xia and El-Azab (2015) employed equation (3.4) in conjunction with a Galerkin-FEM method to model cross-slip of a single dislocation loop or many intensive dislocations on multiple slip systems in an FCC crystal. Zhang and Ngan (2018) used a divergence-preserving finite-volume method to solve equation (3.9) to predict intensive structures including the Frank-Read source and Orowan looping, but this time with full considerations of the long-range elastic and core interactions; see figures (4.5) to (4.8) for some of their results.

In addition to using the exact kinematics laws listed above, other researchers have developed alternative CDD models for intensive GNDs and two such approaches are as follows.

Discrete-continuum model (DCM)—DCM incorporates key elements of DDD into a CDD framework via the use of the finite-element method (FEM). Along this line, Cui *et al* (2022) used the elastoplastic framework introduced in section 5.6 (chapter 5) and cast the velocity gradient in equation (1.1) or (5.51) as:

$$\bar{\mathbf{L}}^{\mathrm{p}} = \sum_k \dot{\gamma}_k \quad \hat{s}_k \otimes \hat{n}_k \tag{7.1}$$

where the shear strain rate field is set to be:

$$\dot{\gamma}_k(\vec{r}) = \oint_{\mathcal{L}^{(k)}(t)} \delta(\vec{r} - \vec{r}') \quad b^{(k)} v^{(k)} \, dr' \tag{7.2}$$

k above denotes a particular dislocation loop of shape $\mathcal{L}^{(k)}(t)$ at time t, and so in equation (7.1), the effects of all dislocation loops in the simulation are summed up. The delta function is effectively the dislocation density for a single dislocation (see equation (3.18)), and hence equation (7.2) is in fact the Orowan–Taylor equation (5.60) for dislocation k. Therefore, instead of using an explicit kinematics law for GND density, line connectivity in the DCM is maintained by using the curves $\mathcal{L}^{(k)}(t)$ as the model parameters; in this sense the DCM is essentially a DDD method. The continuum feature of DCM is to use FEM to directly compute the total stress field from the plastic displacements of all the dislocation curves in the system against the imposed boundary conditions. The procedure is as follows:

- From the dislocation shapes $\mathcal{L}^{(k)}(t)$ at time t, calculate $\bar{\mathbf{L}}^{\mathrm{p}}$ from equations (7.1) and (7.2).
- Update $\mathbf{F}^{\mathrm{p}}$ from the ordinary differential equation $\dot{\mathbf{F}}^{\mathrm{p}} = \bar{\mathbf{L}}^{\mathrm{p}}\mathbf{F}^{\mathrm{p}}$ (equation (5.42)).
- Employ FEM to solve for $\mathbf{F}^{\mathrm{e}}$ from $\mathbf{F} = \mathbf{F}^{\mathrm{e}}\mathbf{F}^{\mathrm{p}}$ (equation (5.40)) using the total deformation gradient $\mathbf{F}$ imposed.
- Calculate the elastic strain rate from equation (5.45) as $\dot{\varepsilon}_{\mathbf{e}} = \mathrm{sym}(\boldsymbol{l}^{\mathrm{e}}) = \frac{1}{2}[\dot{\mathbf{F}}^{\mathrm{e}} \quad \mathbf{F}^{\mathrm{e}-1} + (\dot{\mathbf{F}}^{\mathrm{e}} \quad \mathbf{F}^{\mathrm{e}-1})^{\mathrm{T}}]$, and then elastic strain increment as $\Delta\varepsilon_{\mathbf{e}} = \dot{\varepsilon}_{\mathbf{e}}\Delta t$ for time step Δt.

- Calculate stress increment from Hooke's law (equation (5.64)) as $\Delta\boldsymbol{\sigma} = \mathbb{C}\colon \Delta\boldsymbol{\varepsilon}_{\mathbf{e}}$, then update stress $\boldsymbol{\sigma}(t + \Delta t) = \Delta\boldsymbol{\sigma} + \boldsymbol{\sigma}(t)$.
- Calculate resolved shear stress on dislocation k as: $\tau_{(k)} = \boldsymbol{\sigma}(t + \Delta t)\colon \mathrm{sym}(\hat{s}_k \otimes \hat{n}_k)$, then use dislocation velocity law to update dislocation shapes $\mathcal{L}^{(k)}(t + \Delta t)$.

The DCM therefore makes use of an elastoplastic framework for dynamics closure (section 5.6, chapter 5), and avoids use of the Mura's formula, which is valid only for infinite-sized media.

Dislocation-density potential function (DDPF)—DDPFs are alternative representations of GND densities introduced by Xiang and co-workers (figure 7.1, Zhu and Xiang 2015). One basic DDPF is the $\phi(x, y)$ defined on the slip plane, which is in fact the disregistry function u in the Peierls model (see equation (4.23)). ϕ is related to the GND density ρ_{G} by

$$\rho_{\mathrm{G}} = |\vec{\nabla}\phi| \tag{7.3}$$

and its (exact) kinematics law is:

$$\dot{\phi} = -\rho_{\mathrm{G}} v = -|\vec{\nabla}\phi| v \tag{7.4}$$

where v is the dislocation velocity. Zhu and Xiang (2015) examined in detail how common dislocation features such as line orientation and curvature can be obtained from gradients of DDPFs, and proposed homogenization schemes to reduce 3D dislocation microstructures to DDPF representations.

Summary—Although all CDD models for intensive GNDs surveyed above are based on exact kinematics laws, as discussed in chapter 1, these methods are not valid for mesoscopic simulation of coarse-grained GNDs of extensive dislocation microstructures, where the pixels concerned are necessarily large enough to contain certain SSD contents. As discussed in section 5.1 in chapter 5, the evolution law for coarse-grained GNDs may be written as:

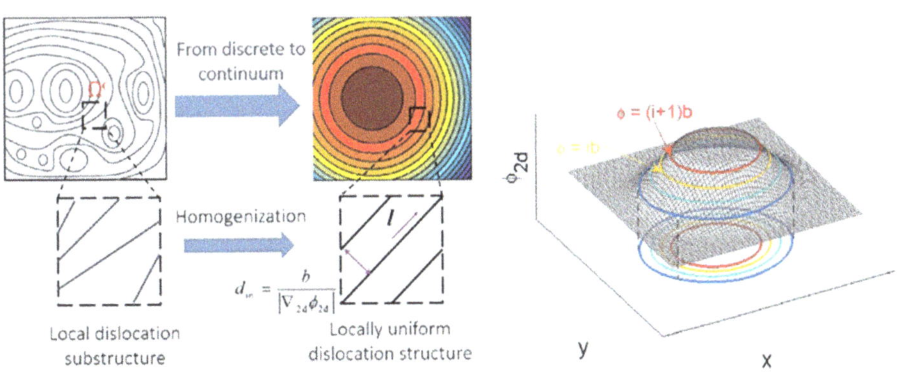

Figure 7.1. The formalism of dislocation-density potential function ϕ for GNDs. (Reproduced from Zhu and Xiang (2015), copyright (2015) with permission from Elsevier.)

$$\dot{\vec{\rho}}_{\mathrm{G}} = -\vec{\nabla} \times (\vec{\rho}_{\mathrm{G}} \times \vec{v}_{\mathrm{G}} + \vec{L}^{\mathrm{p}}) \tag{5.2}$$

but no exact evolution law is known for the SSD flux $\vec{L}^{\mathrm{p}}$. Furthermore, the SSDs can significantly influence the evolution of the coarse-grained GNDs via short-range (Taylor) interactions (see equation (5.23)). As discussed in chapter 1, the main rationale of CDD should be to bridge the mesoscopic gap between DDD and CPFEM methods in figure 1.3 where the dislocation microstructures to be studied should be extensive in nature with significant SSD contents; hence CDD models based solely on coarse-grained GNDs are not useful for this purpose. GND-based CDD models are usable for intensive dislocation microstructures but then this regime is already well covered by DDD or lower models in figure 1.3. All these amount to say that GND-based CDD models, whether in extensive or intensive scales, have limited applicability or usefulness.

7.1.2 Phenomenological CDD for GNDs + SSDs

Next, we review two CDD models that do not have exact kinematics laws for evolution, but nevertheless retain the effects of SSD contents.

Phenomenological mesoscale Field Dislocation Mechanics (PMFDM)—As discussed in section 5.1 in chapter 5, the kinematics law for the GND density can be written as equation (5.2), which can be re-cast for the Nye tensor as:

$$\dot{\boldsymbol{\alpha}} = -\vec{\nabla} \times (\boldsymbol{\alpha} \times \vec{v} + \mathbf{L}^{\mathbf{p}}) \tag{7.5}$$

where $\mathbf{L}^{\mathbf{p}}$ is the deformation gradient due to flux of the SSDs, the exact evolution law of which is unknown. Acharya and Roy (2006) proposed the PMFDM method to use phenomenological laws to model the evolution of $\mathbf{L}^{\mathbf{p}}$. The following is a specific example of PMFDM scheme proposed by Varadhan *et al* (2005), involving a phenomenological Kocks–Mecking type law (equation (1.3)) for the evolution of $\mathbf{L}^{\mathbf{p}}$:

$$\mathbf{L}^{\mathbf{p}} = \sum_{\alpha} \rho_m b v_\alpha \quad \hat{s}_\alpha \otimes \hat{n}_\alpha \tag{7.6}$$

where ρ_m is the density of the mobile SSDs, α denotes the slip system, and other symbols are as defined before. The evolution of ρ_m is assumed to obey a Kocks–Mecking type law:

$$\dot{\rho}_m = \left(\frac{C_1}{b^2} - C_2\rho_m - \frac{C_3}{b}\sqrt{\rho_f}\right)\dot{\Gamma} \tag{7.7}$$

$$\dot{\rho}_f = \left(C_0 b\,|\boldsymbol{\alpha}| + C_2\rho_m + \frac{C_3}{b}\sqrt{\rho_f} - C_4\rho_f\right)\dot{\Gamma} \tag{7.8}$$

Here, ρ_f is the density of the forest obstacles for the mobile SSDs, and $\dot{\Gamma} = |\boldsymbol{\alpha} \times \vec{v} + \mathbf{L}^{\mathbf{p}}|$ is the total equivalent strain rate where $| \cdots |$ denotes the norm of the tensor inside (say, the square root of the sum of the squares of the tensor elements). The various model parameters account for the following:

C_0: contribution of GNDs (represented by $|\boldsymbol{\alpha}|$) to forest hardening;
C_1: multiplication of mobile SSDs;
C_2: mutual annihilation of SSDs to generate debris as forest obstacles;
C_3: immobilization of SSDs on interacting with forest obstacles;
C_4: dynamic recovery of forest obstacles

Sector-wise GND model—To preserve the SSD contents in a CDD formulation, Leung and Ngan (2016) proposed to sum up the densities of individual dislocations residing in a pixel into four GNDs, one for each quadrant of the dislocation character space (figure 7.2):

$$\vec{\rho}_I(\vec{r}) = \rho_I^e \hat{e} + \rho_I^s \hat{s} = \sum_{\substack{i \in \alpha \\ \theta_i \in (0, \pi/2)}} \vec{\rho}_i(\vec{r}) ; \tag{7.9A}$$

$$\vec{\rho}_{II}(\vec{r}) = \rho_{II}^e \hat{e} + \rho_{II}^s \hat{s} = \sum_{\substack{i \in \alpha \\ \theta_i \in (\pi/2, \pi)}} \vec{\rho}_i(\vec{r}) ; \tag{7.9B}$$

$$\vec{\rho}_{III}(\vec{r}) = \rho_{III}^e \hat{e} + \rho_{III}^s \hat{s} = \sum_{\substack{i \in \alpha \\ \theta_i \in (\pi, 3\pi/2)}} \vec{\rho}_i(\vec{r}) ; \tag{7.9C}$$

$$\vec{\rho}_{IV}(\vec{r}) = \rho_{IV}^e \hat{e} + \rho_{IV}^s \hat{s} = \sum_{\substack{i \in \alpha \\ \theta_i \in (3\pi/2, 2\pi)}} \vec{\rho}_i(\vec{r}) \tag{7.9D}$$

Each GND may also be resolved into edge and screw components, ρ_I^e, ρ_I^s, etc. Their evolution laws are analogous to those of GNDs comprising movement and tilting terms (see equation (3.9)), i.e.,

$$\dot{\rho}_I^{e,\text{move}} = -\vec{\nabla} \bullet \left[\rho_I^e \left(\vec{v}_{e+}^\alpha + \vec{v}_{s+}^\alpha\right)\right] + \rho_I^s \left(\frac{\partial v_{s+}^\alpha}{\partial s} + \frac{\partial v_{e+}^\alpha}{\partial e}\right); \tag{7.10A}$$

$$\dot{\rho}_I^{s,\text{move}} = -\vec{\nabla} \bullet \left[\rho_I^s \left(\vec{v}_{e+}^\alpha + \vec{v}_{s+}^\alpha\right)\right] - \rho_I^e \left(\frac{\partial v_{s+}^\alpha}{\partial s} + \frac{\partial v_{e+}^\alpha}{\partial e}\right); \tag{7.10B}$$

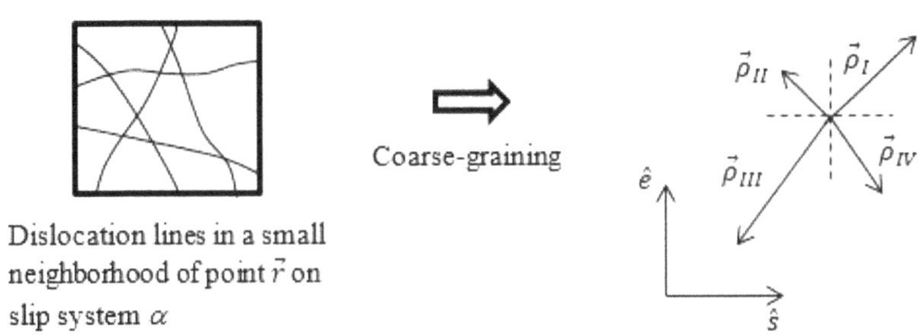

Figure 7.2. Coarse-graining scheme to sum dislocation contents into four GNDs, one for each quadrant of the dislocation-character space. (Reproduced from Leung and Ngan (2016), copyright (2016) with permission from Elsevier.)

$$\dot{\rho}_{II}^{e,\text{move}} = -\vec{\nabla} \bullet \left[\rho_{II}^{e}\left(\vec{v}_{e+}^{\alpha} + \vec{v}_{s-}^{\alpha}\right)\right] + \rho_{II}^{s}\left(\frac{\partial v_{s-}^{\alpha}}{\partial s} - \frac{\partial v_{e+}^{\alpha}}{\partial e}\right); \tag{7.10C}$$

$$\dot{\rho}_{II}^{s,\text{move}} = -\vec{\nabla} \bullet \left[\rho_{II}^{s}\left(\vec{v}_{e+}^{\alpha} + \vec{v}_{s-}^{\alpha}\right)\right] + \rho_{II}^{e}\left(\frac{\partial v_{e+}^{\alpha}}{\partial e} - \frac{\partial v_{s-}^{\alpha}}{\partial s}\right); \tag{7.10D}$$

$$\dot{\rho}_{III}^{e,\text{move}} = -\vec{\nabla} \bullet \left[\rho_{III}^{e}\left(\vec{v}_{e-}^{\alpha} + \vec{v}_{s-}^{\alpha}\right)\right] - \rho_{III}^{s}\left(\frac{\partial v_{s-}^{\alpha}}{\partial s} + \frac{\partial v_{e-}^{\alpha}}{\partial e}\right); \tag{7.10E}$$

$$\dot{\rho}_{III}^{s,\text{move}} = -\vec{\nabla} \bullet \left[\rho_{III}^{s}\left(\vec{v}_{e-}^{\alpha} + \vec{v}_{s-}^{\alpha}\right)\right] + \rho_{III}^{e}\left(\frac{\partial v_{s-}^{\alpha}}{\partial s} + \frac{\partial v_{e-}^{\alpha}}{\partial e}\right); \tag{7.10F}$$

$$\dot{\rho}_{IV}^{e,\text{move}} = -\vec{\nabla} \bullet \left[\rho_{IV}^{e}\left(\vec{v}_{e-}^{\alpha} + \vec{v}_{s+}^{\alpha}\right)\right] + \rho_{IV}^{s}\left(\frac{\partial v_{e-}^{\alpha}}{\partial e} - \frac{\partial v_{s+}^{\alpha}}{\partial s}\right); \tag{7.10G}$$

$$\dot{\rho}_{IV}^{s,\text{move}} = -\vec{\nabla} \bullet \left[\rho_{IV}^{s}\left(\vec{v}_{e-}^{\alpha} + \vec{v}_{s+}^{\alpha}\right)\right] + \rho_{IV}^{e}\left(\frac{\partial v_{s+}^{\alpha}}{\partial s} - \frac{\partial v_{e-}^{\alpha}}{\partial e}\right) \tag{7.10H}$$

Summary—Although these phenomenological CDD models have SSD effects retained, they are not ideal as the kinematics laws are not exact and they do not respect the connectivity of the dislocation lines.

7.2 The 'Beauty and the Beast' of CDD based on ADD

7.2.1 The 'Beauty': kinematics laws

This book, and in particular, chapter 5, has shown that the exact kinematics law is known for the ADD of $\varrho(\vec{r}, \theta)$, i.e.,

$$\dot{\varrho} = -\vec{\nabla}(\varrho v) \bullet \hat{\theta} - \frac{\partial(\varrho\omega)}{\partial\theta} \tag{5.17}$$

Equation (5.17) is exactly what is needed for the development of a satisfactory CCD simulation scheme for the mesoscopic gap in figure 1.3, where the effects of SSDs need to be fully considered. The subsequent scheme presented in chapter 6 is therefore superior to any of the previous CDD schemes surveyed above.

As an example of mathematical beauty in nature, the exact evolution laws for densities of line features discussed so far (equations (3.9) and (5.17)) belong to a family of conservation laws for points, lines and surfaces in the 3D space, as reviewed in chapter 2. Exercise 7.1 below gives the details.

Exercise 7.1. Equations (2.2) and (2.3) give the kinematics law for point density while equation (5.12) gives the kinematics law for line density. By referring to figure 2.2(c) and equation (2.6) for the definition of surface density, derive its kinematics law. Discuss the law's application to grain size in polycrystals.

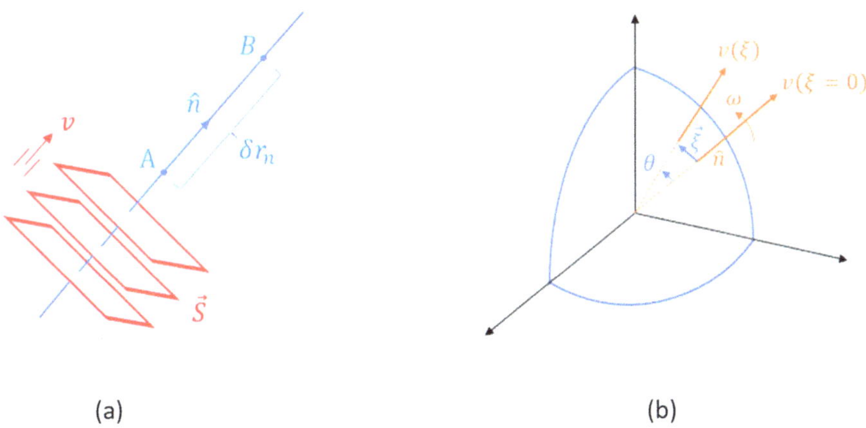

Figure 7.3. (a) Movement of surfaces into a density measurer of line segment AB. (b) The $v(\vec{\xi})$ surface.

Solution: Referring to figure 7.3(a) showing a set of surfaces with density $\vec{S}$ moving (with velocity v) into a density measurer, which is a short line segment AB of infinitesimal length δr_n. Let $\hat{n}$ be the surface normal so that $\vec{S} = S\hat{n}$, S = number of planes (normal to $\hat{n}$) threaded by unit length of line along $\hat{n}$.

During time δt,
number of planes entering AB at A = $vS\delta t$;
number of planes exiting AB at B = $(v + \delta v)(S + \delta S)\delta t = vS\delta t + \frac{\partial(Sv)}{\partial r_n}\delta t \delta r_n$

Therefore, number of planes accumulating in δr_n due to plane movement is:

$$\dot{S}^{\text{move}}\delta t \delta r_n = -\frac{\partial(Sv)}{\partial r_n}\delta t \delta r_n$$

or,

$$\dot{S}^{\text{move}} = -\frac{\partial(Sv)}{\partial r_n} = -\vec{\nabla}(Sv) \bullet \hat{n} \tag{7.11}$$

In addition, in analogy with equation (5.12) for lines, those planes already residing within δr_n may tilt away from the counting line direction $\hat{n}$, leading to a change in $S(\hat{n})$ by the amount:

$$\dot{S}^{\text{tilt}} = -\frac{\partial(S\omega)}{\partial\theta} \tag{7.12}$$

where $\omega = -\partial v/\partial\xi'$; ξ' is along the steepest-descent direction $\vec{\xi}'$ of the $v(\vec{\xi})$ surface (figure 7.3(b)).

The kinematics law for $S(\hat{n})$ is therefore

$$\dot{S}(\hat{n}) = \dot{S}^{\text{move}} + \dot{S}^{\text{tilt}} = -\frac{\partial(Sv)}{\partial r_n} - \frac{\partial(S\omega)}{\partial\theta} \tag{7.13}$$

For a polycrystal, let its grain size in the coarse-grained sense be d, which is a variable of space, time, and the measurer direction $\hat{n}$. Therefore, $d = 1/S$ and hence $\dot{d} = -d^2\dot{S}$. From equation (7.11), movement of the grain boundaries leads to:

$$\dot{d}^{\text{move}} = d^2 \quad \vec{\nabla}\left(\frac{v}{d}\right) \bullet \hat{n} = \underbrace{d(\vec{\nabla}v) \bullet \hat{n}}_{\text{grain growth}} - \underbrace{v(\vec{\nabla}d) \bullet \hat{n}}_{\text{rigid translation}} \tag{7.14}$$

Here, the first term is due to grain growth, since if each grain grows at the same speed about its own centre, a stationary observer will see a spatial gradient of the velocity of the grain boundaries. The second term is due to rigid translation of the entire polycrystalline structure by the velocity v, and this term is non-zero to a stationary observer only if there is a non-zero spatial gradient of the grain size.

From equation (7.12), tilting of the grain boundaries will lead to:

$$\dot{d}^{\text{tilt}} = -d^2 \quad \dot{S}^{\text{tilt}} = \underbrace{-\,\omega\frac{\partial d}{\partial \theta}}_{\text{rotation}} + \underbrace{d\frac{\partial \omega}{\partial \theta}}_{\text{elongation}} \tag{7.15}$$

where the two terms are interpreted as grain boundary rotation and elongation, respectively.

From $\dot{S}(\hat{n}) = \dot{S}^{\text{move}} + \dot{S}^{\text{tilt}}$, we have $\dot{d} = -d^2\dot{S} = \dot{d}^{\text{move}} + \dot{d}^{\text{tilt}}$ and hence the overall evolution equation for d is:

$$\dot{d} = \underbrace{d(\vec{\nabla}v) \bullet \hat{n}}_{\text{grain growth}} - \underbrace{v(\vec{\nabla}d) \bullet \hat{n}}_{\text{rigid translation}} - \underbrace{\omega\frac{\partial d}{\partial \theta}}_{\text{rotation}} + \underbrace{d\frac{\partial \omega}{\partial \theta}}_{\text{elongation}} \tag{7.16}$$

Just like for dislocation density in equation (5.12), dynamics closure needs to be considered to obtain the velocity field v in order to solve equation (7.16), and this is beyond the present scope here.

7.2.2 The 'Beast': dynamics closure

While the exact kinematic law in equation (5.17) for ADD is the 'Beauty' for developing a mesocopic CDD model, dynamics closure needed to solve for the ADD qualifies for the 'Beast'. As in the musical, the Beast is actually a prince to save the Beauty, just that it is ugly. The ugliness in the present context of ADD-based CDD concerns certain limiting or *ad hoc* measures involved in the dynamics closure scheme as well as in the numerical implementation, such as:

- No one-size fitting all dynamics closure scheme—intensive microstructures need core interactions (section 4.3), which are not needed in extensive microstructure simulation.
- Using Mura's formula for the long-range interactions, which is valid only for infinite media.
- The numerical issue of movement-tilting synchronization discussed in section 3.4.3.2 in chapter 3, which could only be solved by the *ad hoc* method of redistributing the density.

- For extensive microstructures, equation (5.17) does not represent dislocation events such as generation by dislocation sources and cross slip. Adding these would involve extra *ad hoc* terms in the kinematics law.

7.3 What requires future work?

One major direction of the development for ADD-based CDD for the mesoscopic scale is to construct capabilities for handling more complex metallurgical features. In addition to adding source and sink terms to the evolution equation to represent dislocation generation and cross-slip as mentioned above, another topic is to model polycrystalline plasticity where ADD interactions with grain boundaries need to be considered. A third topic in this direction is to realistically represent dislocation-precipitate interactions in a mesoscopic coarse-grained manner for two-phase alloys.

Another interesting direction deserving future investigation concerns the very notion of using a finite-sized pixel to represent a reasonable level of coarse graining appropriate to mesoscale simulations in figure 1.3. The kinematic law in equation (5.17) is a field equation without requiring a definition of the pixel scale for the numerical solution, and it is exact as long as one can justify that the dislocation contents are representable by an ADD function that depends on the dislocation character (hence preserving the SSDs). The latter does imply a certain level of coarse graining, since fine-graining down to the atomic scale would not be compatible with the notion of SSDs, as individual dislocations are GNDs when resolved down to the atomic level of graining. Hence, the concept of ADD automatically assumes that graining is at a coarse enough level, and then there is the question of how the coarse-graining level, or selected pixel size, affects the simulation. One key aspect is the intra-pixel Mura interaction—currently, in section 5.5, the long-range (elastic) interactions are considered between pixels, while within a pixel only short-range (Taylor) interactions are assumed. This will be good if the pixel size is small, but for larger pixels (e.g. few tens of b), the elastic interactions within each pixel should constitute a pressure term and hence if there is pressure gradient across the pixel length scale, there should be a density-gradient force driving the density to be more uniform.

Finally, despite the *ad hoc* features of the dynamics closure and the above suggested topics for future investigation, this book has presented a simple account on the basics of CDD, and an appealing approach based on the ADD. As in most subjects in science and engineering, further improvements and developments are always needed, but the basics should allow those interested to get their hands on the subject.

References

Acharya A and Roy A 2006 Size effects and idealized dislocation microstructure at small scales: predictions of a phenonmenological model of mesoscopic field dislocation mechanics: I *J. Mech. Phys. Solids* **54** 1687–710

Cui Y, Wang T, Luo S, Li Z and Li Z 2022 A discrete-continuum model of three-dimensional dislocation elastodynamics. *Int. J. Plast.* **152** 103221

Leung H S and Ngan A H W 2016 Dislocation-density function dynamics—an all-dislocation, full-dynamics approach for modeling intensive dislocation structures *J. Mech. Phys. Solids* **91** 172–203

Varadhan S N, Beaudoin A J, Acharya A and Fressengeas C 2006 Dislocation transport using an explicit Galerkin/least-squares formulation *Model. Simul. Mater. Sci. Eng.* **14** 1245–70

Varadhan S, Beaudoin A J and Freesengeas. C 2005 Coupling the dynamics of statistically distributed and excess dislocations *Int. Conf. on Statistical Mechanics of Plasticity and Related Instabilities* (Trieste: PoS)

Xia S and El-Azab A 2015 Computational modelling of mesoscale dislocation patterning and plastic deformation of single crystals *Model. Simul. Mater. Sci. Eng.* **23** 055009

Zhang Y and Ngan A H W 2018 Dislocation-density dynamics for modeling the cores and Peierls stress of curved dislocations *Int. J. Plast.* **104** 1–22

Zhu Y and Xiang Y 2015 A continuum model for dislocation dynamics in three dimensions using the dislocation density potential functions and its application to micro-pillars *J. Mech. Phys. Solids* **84** 230–53

IOP Publishing

Continuous Dislocation Dynamics Made Simple

Alfonso Ngan and Alireza Kalaei

Appendix A

Basic mathematics for understanding this book

A.1 Tensors in solid mechanics

Tensor algebra is a big topic and the interested reader should refer to a specialized text on this subject. The following is a quick summary to allow the reader to understand the various tensor notations used in this book.

A.1.1 Tensor order and rank

An nth order tensor of rank m, at its simplest, may be understood as a set of numbers $A_{i,\,j,\,k,\,\ldots}$ indexed by n integers i, j, k, … (n of these), each running from 1 to m (the rank):

$$\mathbf{A} = \{A_{i,\,j,\,k,\,\ldots}\} \tag{A.1}$$

There are therefore m^n elements in the tensor. In this book, all the tensors are for quantities defined in the 3D space, and so they are all of rank 3. Tensors are written as bold letters (such as the $\mathbf{A}$ above) or special symbols (such as $\mathbb{C}$ for the elastic constant tensor), and the context should make it clear.

A scalar a is evidently a tensor of 0th order without any need for an index.

A first-order tensor is a vector $\mathbf{v} = \{v_i\}$ where the index i runs from 1 to m (the rank), although in this book we are only interested in vectors of rank 3. In addition to bold letters used generally for tensors, in this book vectors are also specified by an arrow above text, such as $\vec{v}$.

An example of a second-order (third rank) tensor is the stress tensor:

$$\boldsymbol{\sigma} = \{\sigma_{ij}\} = \begin{pmatrix} \sigma_{11} & \sigma_{12} & \sigma_{13} \\ \sigma_{21} & \sigma_{22} & \sigma_{23} \\ \sigma_{31} & \sigma_{32} & \sigma_{33} \end{pmatrix} \tag{A.2}$$

since a stress component σ_{ij} needs to be specified by two indices i and j, which run from 1 to 3 to span the 3D space. The index i here specifies the traction direction while j specifies the normal of the plane on which the traction acts. A stress tensor

doi:10.1088/978-0-7503-6330-3ch8

therefore has $m^n = 3^2 = 9$ elements but, by rotational equilibrium, any stress tensor must also be symmetric, i.e. $\sigma_{ij} = \sigma_{ji}$, or

$$\boldsymbol{\sigma} = \{\sigma_{ij}\} = \boldsymbol{\sigma}^T = \{\sigma_{ji}\} \tag{A.3}$$

In this book, we have encountered only one third-order tensor, namely, the permutation tensor $\{\epsilon_{ijk}\}$ in equations (3.6) and (4.3), where the elements ϵ_{ijk} are 1 when ijk are in cyclic order, -1 when they are in reverse cyclic order, and 0 otherwise. There are altogether $3^3 = 27$ elements in a third-order (rank 3) tensor.

The elastic constants c_{ijkl} in equation (4.22) are indexed by four integers i, j, k and l and so they form a fourth-order (rank 3) tensor $\mathbb{C} = \{c_{ijkl}\}$. There are $3^4 = 81$ elements in a fourth-order (rank 3) tensor, i.e. 81 elastic constants c_{ijkl} for a crystal in general, although crystal symmetry often reduces the independent elastic constants to a low number (3 for isotropic crystal).

A.1.2 Tensors as transformations between tensors

A tensor is necessarily a transformation between other tensors. As odd as this may sound, this statement can be understood easily by the example of the stress tensor $\boldsymbol{\sigma}$ in equations (A.2) and (A.3). By superposition of vectors, the traction in any direction i acting on a plane with normal vector $\mathbf{n} = \{n_j\}$ is given by:

$$T_i = \sigma_{i1}n_1 + \sigma_{i2}n_2 + \sigma_{i3}n_3 = \sum_{j=1}^{3} \sigma_{ij}n_j$$

The repeated index abbreviation—In tensor algebra, it is customary to write the above in a condensed form:

$$T_i = \sigma_{ij}n_j \tag{A.4}$$

where the summation symbol is omitted but summation over the repeated index j is understood.

The three traction components T_i (i= 1–3) in fact form a vector $\mathbf{T} = \{T_i\}$ (a first-order tensor) which can be written as

$$\mathbf{T} = \begin{pmatrix} T_1 \\ T_2 \\ T_3 \end{pmatrix} = \begin{pmatrix} \sigma_{11} & \sigma_{12} & \sigma_{13} \\ \sigma_{21} & \sigma_{22} & \sigma_{23} \\ \sigma_{31} & \sigma_{32} & \sigma_{33} \end{pmatrix} \begin{pmatrix} n_1 \\ n_2 \\ n_3 \end{pmatrix} = \boldsymbol{\sigma} \cdot \mathbf{n} \tag{A.5}$$

or, equivalently,

$$\mathbf{T} = \begin{pmatrix} T_1 & T_2 & T_3 \end{pmatrix} = \begin{pmatrix} n_1 & n_2 & n_3 \end{pmatrix} \begin{pmatrix} \sigma_{11} & \sigma_{12} & \sigma_{13} \\ \sigma_{21} & \sigma_{22} & \sigma_{23} \\ \sigma_{31} & \sigma_{32} & \sigma_{33} \end{pmatrix} = \mathbf{n} \cdot \boldsymbol{\sigma} \tag{A.6}$$

The $\boldsymbol{\sigma}$ in equations (A.5) or (A.6) in fact acts as a transformation that maps a vector ($\mathbf{n}$) to another ($\mathbf{T}$), but then both vectors $\mathbf{n}$ and $\mathbf{T}$ should be independent of the vector basis. This property of a transformation independent of the vector basis is a defining criterion for a second-order tensor, that distinguishes a tensor from a general 3×3 matrix.

Other examples of second-order tensors behaving as transformations in this book include the following:

- Nye tensor (equation (2.10))—transforms the normal of a unit plane to the total Burgers vector of dislocations threading the plane;
- Schmid tensor (equation (5.32))—transforms a line segment $d\overline{\mathbf{X}}$ to a vector along the slip direction with length equal to the projection of the line segment on the slip plane normal (see equation (5.48) and figure 5.8));
- deformation gradient **F** (equation (5.35))—transforms a line segment $d\mathbf{X}$ in an undeformed crystal to the corresponding material line segment $d\mathbf{x}$ in the deformed crystal (see equation (3.35)).

For the fourth-order elastic constant tensor $\mathbb{C} = \{c_{ijkl}\}$, Hooke's law (equation (5.69)) is written as

$$\sigma_{ij} = c_{ijkl}\varepsilon_{e_{kl}} \text{ (summation over repeated indices } k \text{and } l) \tag{A.7}$$

which can be written as a double dot product (:) between the tensors of stress $\boldsymbol{\sigma}$ and elastic strain $\boldsymbol{\varepsilon}_{\mathbf{e}}$:

$$\boldsymbol{\sigma} = \mathbb{C} : \boldsymbol{\varepsilon}_{\mathbf{e}} \tag{A.8}$$

Again, $\mathbb{C}$ here acts as a transformation, but this time, between two second-order tensors $\boldsymbol{\sigma}$ and $\boldsymbol{\varepsilon}_{\mathbf{e}}$. The double dot product is explained below.

A.1.3 Tensors from dyadic multiplication

Of the various examples of tensors considered above, the Nye and Schmid tensors (equations (2.10A) and (5.32)) are defined by *dyadic multiplication* between two vectors:

$$\text{Nye tensor: } \boldsymbol{\alpha} = \vec{b} \otimes \vec{\rho} \tag{A.9}$$

$$\text{Schmid tensor: } \hat{s}_\alpha \otimes \hat{n}_\alpha \tag{A.10}$$

Take the notation of the Nye tensor as illustration (equation (2.10B)), the dyadic means that

$$\boldsymbol{\alpha} = \{\alpha_{ij}\} = \begin{pmatrix} \alpha_{11} & \alpha_{12} & \alpha_{13} \\ \alpha_{21} & \alpha_{22} & \alpha_{23} \\ \alpha_{31} & \alpha_{32} & \alpha_{33} \end{pmatrix} = \{b_i\rho_j\} = \begin{pmatrix} b_1\rho_1 & b_1\rho_2 & b_1\rho_3 \\ b_2\rho_1 & b_2\rho_2 & b_2\rho_3 \\ b_3\rho_1 & b_3\rho_2 & b_3\rho_3 \end{pmatrix} \tag{A.11}$$

Not all second-order tensors can be composed meaningfully by dyadic multiplication, say, the stress tensor $\boldsymbol{\sigma}$ and deformation gradient tensor **F** discussed above are not formed by dyadic multiplication between two meaningful vectors.

Exercise A.1 Explain from equation (A.11) why the Nye tensor is the total Burgers vector of dislocations threading a unit plane of a given normal direction.

Solution: Let $\mathbf{n} = \{n_j\}$ be the normal direction of the unit area. From equation (A.11),

$$\boldsymbol{\alpha} \cdot \mathbf{n} = \{b_i \rho_j n_j\} = \{b_i (\boldsymbol{\rho} \cdot \mathbf{n})\}$$

Here, the dot product $\boldsymbol{\rho} \cdot \mathbf{n}$ is simply the number of dislocations threading the unit area; thus $\boldsymbol{\alpha} \cdot \mathbf{n}$ is a vector equal to the total Burgers vector of the dislocations threading the unit area.

A.1.4 Tensor products

In addition to the dyadic product explained in section A.1.3, the following products between tensors and vectors are used in this book.

Inner (or dot) product (•)—In a dot product, the output tensor has elements that are products of the elements of the multiplying tensors involving *single summation of one repeated index*. An example concerns equations (A.4) and (A.5) in which the product $\mathbf{T} = \boldsymbol{\sigma} \cdot \mathbf{n}$ or $\mathbf{T} = \mathbf{n} \cdot \boldsymbol{\sigma}$ has elements $T_i = \sigma_{ij} n_j$, where the summation is over the repeated index j. In this case, the dot product is between a second-order tensor ($\boldsymbol{\sigma}$) and a first-order tensor ($\mathbf{n}$), and the outcome is a first-order tensor ($\mathbf{T}$).

As another example, the decomposition of deformation gradient $\mathbf{F}$ into plastic and elastic components in equation (5.40) is also a dot product:

$$\mathbf{F} = \mathbf{F}^{\mathbf{e}} \cdot \mathbf{F}^{\mathbf{p}} \text{ or } F_{ij} = F^e_{ik} F^p_{kj} \text{ (summation over repeated index } k\text{)} \tag{A.12}$$

Here, the two multiplying tensors $\mathbf{F}^{\mathbf{p}}$ and $\mathbf{F}^{\mathbf{e}}$ are second order, and the outcome $\mathbf{F}$ is also second order. The multiplication here is the same as simple multiplication of two 3×3 matrices returning a 3×3 matrix as the product.

Finally, as in the $\vec{\nabla} \cdot \vec{J}$ in equation (2.2) and the $(\hat{n}_\alpha \cdot \hat{e})$ in equation (5.48), the simple dot product between two vectors (1st order tensors) is a scalar (0th order tensor). For vector multiplication, the simple dot product is remarkedly different and not to be confused with the dyadic product in section A.1.3, which returns a second-order tensor instead of a scalar.

In all cases above, when there is no ambiguity, the dot (•) may be omitted in writing.

Double inner (or dot) product (:), *aka tensor contraction*—In a double dot product, the output tensor has elements that are products of the elements of the multiplying tensors involving *double summation of two repeated indices*. Let us examine two examples.

First, in Hooke's law $\boldsymbol{\sigma} = \mathbb{C}\colon \boldsymbol{\varepsilon}_{\mathbf{e}}$ in equation (A.8), the product $\boldsymbol{\sigma} = \{\sigma_{ij}\}$ has elements given by $\sigma_{ij} = c_{ijkl}\varepsilon_{ekl}$ (equation (A.6)), in which double summation is over the two repeated indices of k and l. Here, the double dot product is between a fourth-order tensor ($\mathbb{C} = \{c_{ijkl}\}$) and a second-order tensor ($\boldsymbol{\varepsilon}_{\mathbf{e}} = \{\varepsilon_{ekl}\}$), and the outcome is a second-order tensor ($\boldsymbol{\sigma} = \{\sigma_{ij}\}$).

As a second example, the calculation of the resolved shear stress $\tau_{\alpha}^{\text{app}}$ in equation (5.70) also involves a double dot product:

$$\tau_{\alpha}^{\text{app}} = \boldsymbol{\sigma}\text{: sym}\left(\hat{s}_{\alpha} \otimes \hat{n}_{\alpha}\right) = \frac{1}{2}\sigma_{ij}(s_i n_j + s_j n_i) \text{ (summation over } i \text{ and } j) \qquad \text{(A.13)}$$

Here, the two multiplying tensors $\boldsymbol{\sigma}$ and sym$(\hat{s}_{\alpha} \otimes \hat{n}_{\alpha})$ are second order, and the outcome is a scalar (0th order tensor).

Cross product—Cross products between a second-order tensor $\boldsymbol{T}$ and a vector $\vec{u}$ are defined as follows (see equations (3.6A) and (3.6B)):

$$\boldsymbol{T} \times \vec{u} = \{T_{ij} u_k \epsilon_{jkl}\} = \text{a 2nd order tensor } (3 \times 3 \text{ matrix}) \text{ with each row } i \text{ being the simple vector cross product of (the same row } i \text{ of } \boldsymbol{T}) \times \vec{u}; \qquad \text{(A.14)}$$

$$\vec{u} \times \boldsymbol{T} = \{u_k T_{ij} \epsilon_{jkl}\} = \text{a 2nd order tensor } (3 \times 3 \text{ matrix}) \text{ with each row } i \text{ being the simple vector cross product of } \vec{u} \times \text{ (the same row } i \text{ of } \boldsymbol{T}). \qquad \text{(A.15)}$$

Here, $\{\epsilon_{jkl}\}$ is the permutation tensor discussed above. Note that in both equations above, double summation takes place over the repeated indices j and k, and then the element $T_{ij} u_k \epsilon_{jkl}$ or $u_k T_{ij} \epsilon_{jkl}$ is indexed by i and l in a second-order tensor or 3×3 matrix.

Identity tensor and inverse—In this book, we only encounter the identity tensor and inverses for second-order tensors, so we will limit to second order. The second-order identity tensor **I** has the property that its dot product with any second-order tensor **A** is just **A** itself:

$$\mathbf{I} \cdot \mathbf{A} = \mathbf{A} \qquad \text{(A.16)}$$

Obviously,

$$\mathbf{I} = \{\delta_{ij}\} = \begin{pmatrix} 1 & 0 & 0 \\ 0 & 1 & 0 \\ 0 & 0 & 1 \end{pmatrix} \qquad \text{(A.17)}$$

The inverse of **A**, written as $\mathbf{A}^{-1}$, is the tensor such that

$$\mathbf{A} \cdot \mathbf{A}^{-1} = \mathbf{I} \qquad \text{(A.18)}$$

A.2 Calculus of scalar, vector and tensor fields

A scalar, vector or tensor field is a function of space where the functional value is a scalar (0th order tensor), vector (1st order tensor), or tensor (order 2 or above), respectively. The chemical potential μ in equation (2.3) is an example of a scalar field:

$$\vec{J} = -\frac{D}{kT} C\, \vec{\nabla} \mu \qquad \text{(A.19)}$$

where μ here is understood to be a function of space, i.e. $\mu(\vec{r})$ ($\vec{r}$ = position vector in space), and as a chemical potential (the energy change when a single defect is added to point $\vec{r}$), μ is a scalar.

The $\vec{J}$ in equation (A.19) is a vector field, since flux $\vec{J}$ is a vector and it is changing over space, i.e. it is a function $\vec{J}(\vec{r})$.

An example of a tensor field is the stress $\boldsymbol{\sigma} = \{\sigma_{ij}\}$ in equation (A.2), where each element is a scalar function of space, i.e. $\sigma_{ij}(\vec{r})$.

Note that in the examples above the scalar, vector or tensor can also be varying with time, so they may also be functions of time in addition to space.

The del operator—Spatial derivatives of scalar, vector or tensor fields are usually expressed in formats involving the gradient or 'del' operator $\vec{\nabla}$, which is defined in equation (2.2) as:

$$\vec{\nabla} = \hat{e}_1\left(\frac{\partial}{\partial x_1}\right) + \hat{e}_2\left(\frac{\partial}{\partial x_2}\right) + \hat{e}_3\left(\frac{\partial}{\partial x_3}\right) = \hat{e}_i\left(\frac{\partial}{\partial x_i}\right) \tag{A.20}$$

Here, $\hat{e}_1$, $\hat{e}_2$ and $\hat{e}_3$ are the unit vectors in the Cartesian basis set of the space, and x_1, x_2 and x_3 are coordinates of $\vec{r}$ along $\hat{e}_1$, $\hat{e}_2$ and $\hat{e}_3$, respectively. When there is ambiguity over the coordinates, such as the case of the elastoplastic framework in figure 5.6 where the vector $\mathbf{X}$ (or $\vec{X}$) marks the material coordinates in the undeformed body while $\mathbf{x}$ (or $\vec{x}$) marks the material coordinates in the deformed body in the same basis set, and there is a need to distinguish between the two, then the del symbol may be written as:

$$\vec{\nabla}_{\mathbf{X}} = \hat{e}_i\left(\frac{\partial}{\partial X_i}\right), \quad \vec{\nabla}_{\mathbf{x}} = \hat{e}_i\left(\frac{\partial}{\partial x_i}\right)$$

where $\hat{e}_i$ (i = 1–3) are the common basis-set unit vectors, and X_i and x_i are, respectively, the coordinates of $\mathbf{X}$ and $\mathbf{x}$ in the same basis set. Such notations for 'del' are used in exercise 5.6 in chapter 5.

There is a further short-hand notation for the partial derivatives in the above: $\partial/\partial x_i$ can be abbreviated as $\partial_{,i}$, and then $\vec{\nabla} = \hat{e}_i\,\partial_{,i}$.

The operations of 'del' on scalar, vector and tensor fields are discussed below.

Gradient of scalar fields—When operated on a scalar field such as the $\vec{\nabla}\mu$ in equation (A.19), the outcome:

$$\vec{\nabla}\mu = \hat{e}_i\left(\frac{\partial \mu}{\partial x_i}\right) \text{ or } \vec{\nabla}\mu = \hat{e}_i \mu_{,i} \tag{A.21}$$

is the *gradient vector* of the $\mu(\vec{r})$ landscape in the 3D space. The (scalar) gradient of the $\mu(\vec{r})$ landscape along a given direction $\hat{\xi}$, defined as $\partial\mu/\partial\xi$ where ξ is the coordinate along $\hat{\xi}$, is given by the chain rule as:

$$\frac{\partial \mu}{\partial \xi} = \frac{\partial \mu}{\partial x_i}\frac{\partial x_i}{\partial \xi} = \vec{\nabla}\mu \cdot \frac{\partial(\xi\hat{\xi})}{\partial \xi} = \vec{\nabla}\mu \cdot \hat{\xi} \tag{A.22}$$

This book has many examples of this type of gradient, such as the $\vec{\nabla}(\rho v) \cdot \hat{v}$ term in equation (3.9).

Laplacian operator—The Laplacian operator is the dot product of two del operators:

$$\nabla^2 = \vec{\nabla} \cdot \vec{\nabla} = \left[\hat{e}_i\left(\frac{\partial}{\partial x_i}\right)\right] \cdot \left[\hat{e}_i\left(\frac{\partial}{\partial x_i}\right)\right] = \frac{\partial^2}{\partial x_1^2} + \frac{\partial^2}{\partial x_2^2} + \frac{\partial^2}{\partial x_3^2} \tag{A.23}$$

∇^2 is operable on scalar, vector or tensor fields, and the $\nabla^2 C$ in equation (2.5) is an example for a scalar field:

$$\nabla^2 C = \frac{\partial^2 C}{\partial x_1^2} + \frac{\partial^2 C}{\partial x_2^2} + \frac{\partial^2 C}{\partial x_3^2} \tag{A.24}$$

Divergence of vector fields—When the del operates on a vector field, since del itself is a vector, the first possibility is a dot product between the del and the vector field. An example is the $\vec{\nabla} \cdot \vec{J}$ in equation (2.2), which means:

$$\vec{\nabla} \cdot \vec{J} = \left(\hat{e}_1\frac{\partial}{\partial x_i}\right) \cdot (J_i \hat{e}_i) = \frac{\partial J_i}{\partial x_i} = J_{i,j} \tag{A.25}$$

Here, the result is a scalar known as the *divergence*. When the divergence is positive, the vectors of the field $\vec{J}$ point away from the location under consideration, and when it is negative, they point towards the location. The divergence of flux often appears in conservative laws as positive flux divergence means net material movement out of the location and hence depletion at the location, and vice versa for negative flux divergence. The divergence does not need to be always cast in Cartesian coordinates in the form of equation (A.25); it can also be cast in cylindrical coordinates as in equation (3.10).

Curl of vector fields—The other possibility of operating del on a vector field $\vec{u}$ is the 'curl', which is the cross product between del and the vector field $\vec{u} = \hat{e}_i u_i$, i.e.

$$\begin{aligned}\operatorname{curl} \vec{u} = \vec{\nabla} \times \vec{u} &= \begin{pmatrix} \hat{e}_1 & \hat{e}_2 & \hat{e}_3 \\ \dfrac{\partial}{\partial x_1} & \dfrac{\partial}{\partial x_2} & \dfrac{\partial}{\partial x_3} \\ u_1 & u_2 & u_3 \end{pmatrix} \\ &= \hat{e}_1\left(\frac{\partial u_3}{\partial x_2} - \frac{\partial u_2}{\partial x_3}\right) + \hat{e}_2\left(\frac{\partial u_1}{\partial x_3} - \frac{\partial u_3}{\partial x_1}\right) + \hat{e}_3\left(\frac{\partial u_2}{\partial x_1} - \frac{\partial u_1}{\partial x_2}\right)\end{aligned} \tag{A.26}$$

which can also be written in the condensed forms of

$$\operatorname{curl} \vec{u} = \vec{\nabla} \times \vec{u} = \epsilon_{ijk} \hat{e}_i \frac{\partial u_k}{\partial x_j} = \epsilon_{ijk} \hat{e}_i u_{k,j} \tag{A.27}$$

Examples are encountered in equation (3.4) and exercise 3.2. The curl is itself a vector, with component $(\partial u_k/\partial x_j - \partial u_j/\partial x_k)$ along $\hat{e}_i$ $(i \neq j \neq k)$ that marks the rotation of the vector field about the $\hat{e}_i$ axis.

Curl of tensor fields—In this book, the curl operator is also applied to second-order tensors including the Nye tensor (equation (3.5)) and deformation gradient tensors (equation (5.61)). In these cases, the definition in equation (A.15) applies, i.e. for a second-order tensor $\mathbf{A} = \{A_{ij}\}$,

$$\vec{\nabla} \times \mathbf{A} = \{\partial_{,k} A_{ij} \epsilon_{jkl}\} \tag{A.28}$$

For instance, in the $\vec{\nabla} \times (\boldsymbol{\alpha} \times \vec{v})$ in equation (3.5) where $\boldsymbol{\alpha}$ is the second-order Nye tensor and $\vec{v}$ the velocity vector, by virtue of the definition in equation (A.14), $\mathbf{A} = \boldsymbol{\alpha} \times \vec{v} = \{A_{il} = \alpha_{ij} v_k \epsilon_{jkl}\}$ is a second-order tensor and then by equation (A.28) above, $\mathbf{B} = \vec{\nabla} \times (\boldsymbol{\alpha} \times \vec{v}) = \{B_{in} = \partial_{,m} A_{il} \epsilon_{lmn}\}$ is another second-order tensor with elements $B_{in} = \partial_{,m} \alpha_{ij} v_k \epsilon_{jkl} \epsilon_{lmn}$.

Stoke's and Gauss' theorem—For a one-dimensional function $f(x)$, we have a simple identity $\int_a^b f'(x')\mathrm{d}x' = f(b) - f(a)$ that relates an integral of the derivative $f'(x) = \mathrm{d}f(x)/\mathrm{d}x$ over a domain $x \in [a, b]$ to the functional values $f(a)$ and $f(a)$ at the boundaries $x = a$, $x = b$ of the domain. For a vector or tensor function defined in 3D space, Stoke's theorem gives a corresponding identity that equates the *surface integral* of the 'derivative' (i.e. curl) of the function over a surface domain S, to the *contour integral* of the function along the boundary C of the domain. The proof of Stoke's theorem follows directly from the definition of the curl in equations (A.26) and (A.27) and is given in elementary texts of engineering mathematics; hence instead of showing the proof here, the theorem is simply quoted as:

$$\oint_C \vec{v} \cdot d\vec{r} = \iint_S (\vec{\nabla} \times \vec{v}) \cdot \hat{n}\, \mathrm{d}S \tag{A.29}$$

for a vector field $\vec{v}$, and

$$\oint_C \mathbf{A} \cdot \mathrm{d}\vec{r} = \iint_S (\vec{\nabla} \times \mathbf{A}) \cdot \hat{n}\, \mathrm{d}S \tag{A.30}$$

for a tensor field $\mathbf{A}$, where $\hat{n}$ is the unit normal of infinitesimal area dS within the surface S.

Moving one dimension up to volume integral, there is also the Gauss theorem, which equates the *volume integral* of the divergence of a vector field $\vec{v}$ over a volume V to the *surface integral* of the vector field at the boundary surface S of the volume:

$$\iiint_V (\vec{\nabla} \cdot \vec{v})\mathrm{d}V = \oiint_S (\vec{v} \cdot \hat{n})\mathrm{d}S \tag{A.31}$$

where $\hat{n}$ is the normal to the surface S. The reader will recognize that the Gauss theorem is the basis of the conservative law for diffusion in equation (A.25).

www.ingramcontent.com/pod-product-compliance
Ingram Content Group UK Ltd.
Pitfield, Milton Keynes, MK11 3LW, UK
UKHW051850100926
8556IPUK00005B/557